1. Drill out the box and attach the hardware.

2. Solder jumper wires and install the voltage regulator.

3. Twist the rectifier and capacitor leads together, and solder.

4. Crimp the connections and close the box.

5. Mount the box on the bike.

6. Connect the generator.

With a few parts, some quality tools, and a bit of know-how, it's easy to build this simple circuit and breathe new life into that old bottle alternator. Pump up your power supply with your pedals!

Parts list:

- 5V voltage regulator (not pictured)
- Bridge rectifier (not pictured)

Project enclosure 3"×2"×1"

Adaptaplug™ socket cable

1,000µF capacitor

Micro-USB Adaptaplug™

Ring-tongue crimp lugs

Round-head machine screw

Tools checklist:

- Drill with ¼" bit
- Mini long-nose pliers
- Gauged wire stripper/cutter
- 4-way crimping tool
- Soldering iron and solder

D1308594

For complete instructions and details on this project visit:

radioshackdiy.com/pedal-power-phone-charger

radioshack

Make:Volume 36

PUBLISHER'S NOTE - NEW IN 2014

Volume 36 completes the 9th year of MAKE magazine. Based on your feedback, we're happy to announce that starting in 2014, MAKE will be published SIX times a year. You'll get two more volumes at the current annual subscription price. In addition, MAKE will be published in a new larger format, allowing us to cover more projects in an even more visually compelling way. Thank you for supporting MAKE and being a part of the maker community. We are looking forward to a bigger and better 2014. — DD

66

▶▶ PINOCCIO:
Capitalizing on a probletunity.

ON THE COVER

Board Games: Loads of new microcontrollers and single-board computers vie for your attention. Photography by Gunther Kirsch

46

PICK YOUR BRAIN

Vol. 36, Oct 2013. MAKE (ISSN 1556-2336) is published bi-monthly by Maker Media, Inc. in the months of February, April, June, August, October, and December. Maker Media is located at 1005 Gravenstein Hwy. North, Sebastopol, CA 95472, (707) 827-7000. SUBSCRIPTIONS: Send all subscription requests to MAKE. P.O. Box 17046, North Hollywood, CA 91615-9588 or subscribe online at makezine.com/offer or via phone at (866) 289-8847 (U.S. and Canada); all other countries call (818) 487-2037. Subscriptions are available for $34.95 for 1 year (6 issues) in the United States; in Canada: $39.95 USD; all other countries: $49.95 USD. Periodicals Postage Paid at Sebastopol, CA, and at additional mailing offices. POSTMASTER: Send address changes to MAKE, P.O. Box 17046, North Hollywood, CA 91615-9588. Canada Post Publications Mail Agreement Number 41129568. CANADA POSTMASTER: Send address changes to: Maker Media, PO Box 456, Niagara Falls, ON L2E 6V2

Twice the suction of any other vacuum.*

160			
68			
43			
0			

43 Air Watts   **68** Air Watts   **160** Air Watts

Suction power is measured in Air Watts and, as you can see, Dyson DC41 has twice as much suction at the cleaner head than any other vacuum.* Its cleaner head self-adjusts to seal in that suction across carpets and hard floors. And, its Dyson cyclone technology captures more microscopic dust than any other.

dyson.com/suction

dyson

Make: Volume 36

READ ME Always check makezine.com/36 before you get started on projects. There may be important updates or corrections.

PROJECTS

Squeezing Atoms: Build a purple-plasma particle accelerator and explore nuclear fusion on your workbench. **90**

Little Labyrinth: Transform an ordinary carpet into a miniature hedge maze. **140**

127
《 Project(ion):
Share your smartphone photos on the big screen.

132
》 Feel the Fizz:
Ferment a batch of grown-up juice.

> "It's still magic, even if you know how it's done."
> —Terry Pratchett, "A Hat Full of Sky"

Make:

FOUNDER & PUBLISHER
Dale Dougherty
dale@makezine.com

EDITOR-IN-CHIEF
Mark Frauenfelder
markf@makezine.com

VICE PRESIDENT
Sherry Huss
sherry@makezine.com

EDITORIAL

EXECUTIVE EDITOR
Mike Senese
msenese@makezine.com

EDITORIAL DIRECTOR
Ken Denmead
kdenmead@makezine.com

MANAGING EDITOR
Cindy Lum
clum@makezine.com

PROJECTS EDITOR
Keith Hammond
khammond@makezine.com

SENIOR EDITOR
Goli Mohammadi
goli@makezine.com

SENIOR EDITOR
Stett Holbrook
sholbrook@makezine.com

TECHNICAL EDITOR
Sean Michael Ragan
sragan@makezine.com

ASSISTANT EDITOR
Laura Cochrane

STAFF EDITOR
Arwen O'Reilly Griffith

EDITORIAL ASSISTANT
Craig Couden

COPY EDITOR
Laurie Barton

SENIOR EDITOR, BOOKS
Brian Jepson

EDITOR, BOOKS
Patrick DiJusto

DESIGN, PHOTOGRAPHY & VIDEO

CREATIVE DIRECTOR
Jason Babler
jbabler@makezine.com

SENIOR DESIGNER
Juliann Brown

SENIOR DESIGNER
Pete Ivey

ASSOCIATE PHOTO EDITOR
Gregory Hayes

VIDEOGRAPHER
Nat Wilson-Heckathorn

FABRICATOR
Daniel Spangler

WEBSITE

WEB DEVELOPER
Jake Spurlock
jspurlock@makezine.com

WEB DEVELOPER
Cole Geissinger

WEB PRODUCER
Bill Olson

CUSTOMER SERVICE

CUSTOMER CARE TEAM LEADER
Daniel Randolph
cs@readerservices.
makezine.com

SALES & ADVERTISING

SENIOR SALES MANAGER
Katie D. Kunde
katie@makezine.com

SALES MANAGER
Cecily Benzon
cbenzon@makezine.com

SALES MANAGER
Brigitte Kunde
brigitte@makezine.com

CLIENT SERVICES MANAGER
Miranda Mager

CLIENT SERVICES MANAGER
Mara Lincoln

EXECUTIVE ASSISTANT
Suzanne Huston

FINANCE CONTROLLER
Kevin Gushue

COMMERCE

VICE PRESIDENT OF COMMERCE
David Watta

DIRECTOR, RETAIL MARKETING & OPERATIONS
Heather Harmon Cochran
heatherh@makezine.com

MAKER SHED GRAPHIC DESIGNER
Uyen Cao

OPERATIONS MANAGER
Rob Bullington

MARKETING

SENIOR DIRECTOR OF MARKETING
Vickie Welch
vwelch@makezine.com

MARKETING COORDINATOR
Meg Mason

MARKETING COORDINATOR
Karlee Vincent

MARKETING ASSISTANT
Courtney Lentz

MAKER FAIRE

PRODUCER
Louise Glasgow

MARKETING & PR
Bridgette Vanderlaan

PROGRAM DIRECTOR
Sabrina Merlo

BUSINESS DEVELOPMENT MANAGER
Heather Brundage

CHANNEL MANAGER
Kaitlyn Amundsen

PRODUCT DEVELOPMENT ENGINEER
Eric Weinhoffer

MAKER SHED EVANGELIST
Michael Castor

PUBLISHED BY
MAKER MEDIA, INC.
Dale Dougherty, CEO

Copyright © 2013
Maker Media, Inc.
All rights reserved.
Reproduction without
permission is prohibited.
Printed in the USA by
Schumann Printers, Inc.
Visit us online:
makezine.com
Comments may be sent to:
editor@makezine.com

**Manage your account online,
including change of address:**
makezine.com/account
866-289-8847 toll-free
in U.S. and Canada
818-487-2037,
5 a.m.–5 p.m., PST
Follow us on Twitter:
@make   @makerfaire
@craft   @makershed
On Google+: google.com/+make
On Facebook: makemagazine

CONTRIBUTING EDITORS
William Gurstelle, Charles Platt, Matt Richardson

CONTRIBUTING WRITERS
John Abella , Massimo Banzi, Jacob Beningo, Sally Carson,
Chris Connors, David Cook, Limor Fried, Albert den Haan,
Lonnie Honeycutt, Tim Hunkin, Jon Kalish, Bob Knetzger,
Tod E. Kurt, Mike Kuniavsky, Taylor Levy, Forrest M. Mims III,
Syuzi Pakhchyan, Tom Parker, Photojojo, Trevor Shannon,
Dr. Nevin J. Stewart, Werner Strama, Phil Torrone,
Marc de Vinck, Shawn Wallace

CONTRIBUTING ARTISTS:
Fon Davis, Nate Van Dyke, Tim Hunkin, Bob Knetzger,
Damien Scogin, Julie West, Shannon Wheeler, Book Williams Jr.

CONTRIBUTING DESIGNERS:
James Burke, Boni Uzilevsky

ONLINE CONTRIBUTORS
Alasdair Allan, John Baichtal, Meg Allan Cole, Michael Colombo,
Jimmy DiResta, Nick Normal, Haley Pierson-Cox,
Andrew Salomone, Andrew Terranova, Glen Whitney

TECHNICAL ADVISORY BOARD
Kipp Bradford, Evil Mad Scientist Laboratories, Limor Fried,
Joe Grand, Saul Griffith, Bunnie Huang, Tom Igoe, Steve Lodefink,
Erica Sadun, Marc de Vinck

INTERNS
Kelley Benck (engr.), Eric Chu (engr.), Paloma Fautley (engr.),
Sam Freeman (engr.), Andrew Katz (jr. engr.) Gunther Kirsch (photo),
Raghid Mardini (engr.), Brian Melani (engr.), Nick Parks (engr.),
Eloy Salinas (engr.)

PRINTED WITH SOY INK

CONTRIBUTORS

At MAKE, we're lucky enough to have some very talented engineering interns to test our projects. **Raghid Mardini** is a mechanical engineering student at University of California, Berkeley. He enjoys traveling, camping, learning about other cultures, and searching for the best sushi. His life goal is to eventually open a robotics institute in his home country, Syria, because he believes in technology's ability to change people's lives. **Kelley Benck** took the Project Make class at Analy High School in Sebastopol, Calif., and started interning last spring. She "was nervous about what sort of projects I would be asked to do, but those jitters were shaken away when I was handed the daunting Tesla Coil." She hopes to design and 3D print a model 1966 Austin Healey. **Eloy Salinas** has been taking things apart since he was 6; "I broke my first computer at 8 trying to see how the hard drive works." After a bicycling accident in 7th grade left him homebound, he started "programming and hacking everything." When not interning at MAKE, he is a freelance web developer who lives in Santa Cruz, Calif. with his yellow lab.

Albert den Haan (*Capstan Kite Winder*) started as a farm boy and studied computer science for air conditioning. He has shuttled across Canada in search of the perfect workshop to build, alternately the most complicated and the simplest things he can think of. Electro-mechanics blurs that spectrum while pulling cheap aeronautics into the practical. He likes to say, "The second one of anything is harder to build as well or as quickly as the first," so he makes each attempt another first item: different but related. Albert suffers from P.E.B.L.E.: Projects Envisioned Beyond Life Expectancy.

Taylor Levy (*Kickscooter Kickstarter*) is both "serious and smiley." She makes things in her Brooklyn, N.Y., basement with her husband, collaborator, and "partner in all things," Che-Wei, and this fall they're heading off to the MIT Media Lab. Right now they're working on a "really cool backpack that is super-close to launching"; their free time is spent working on their house. "It is mostly fun, but when it isn't, we eat ice cream," she says. Her favorite food is a green smoothie, her favorite color is fluorescent orange, and her favorite tool is a caliper.

After a long stint as a graphic designer, **Book Williams Jr.** (Boards section opener art) rented a small dark basement from a gallery, locked himself in, and experimented with paper for more than a year. He emerged with a unique, paper-crafty illustration style and scraps of card stock in his teeth. Born and raised in Chicago, he now resides in Denver with his girlfriend and wonders why mountains and camping are such a big deal. He digs comics. He digs his bone folder, and he hasn't driven a car in more than 10 years.

Sally Carson (*The Tale of Pinoccio*) is a cartoonist, a retired bike messenger, and the CEO of Pinoccio. She is a passionate bicyclist ("Bikes! Bikes! Bikes!"), and she and her husband Tommy live car-free in Ann Arbor, Mich., where they put studded tires on their bikes so they can ride through the winter. (Other bike enthusiasts can download a comic she made about fixing a flat at makezine.com/go/fixaflat.) She keeps bees, thinks about social insects and mesh networking, fixes bikes, draws comics, and feeds snacks to the chipmunk that lives under her porch.

By Dale Dougherty

Computers in the Mist

Carl Helmers was designing spaceships in kindergarten. He "lucked out" by learning computers in high school in New Jersey, where he eventually got a summer job programming at Bell Labs. Then, as a NASA contractor in Houston, he installed compilers and even wrote a landing program for the Apollo Lunar Module.

Computers were big, expensive machines in the '70s. At an Intel press introduction for the 4004 and 8008 microprocessors, Helmers realized he could now afford to build one from off-the-shelf parts: "A lot of guys like me who had experience working for other people with computers began building computers of our own."

Just what were these small computers good for? Hobbyists were searching for answers, so Helmers created a magazine for them called Byte. In its first issue in September 1975, Helmers wrote that for the hardware person "the fun is in the building," not using or programming. "The software is an exploration of the possibilities of the hardware." But the whole point of homebrew computers was "to come up with interesting and exotic applications." A computer experimenter was looking up at a large, unclimbed mountain with three possible ascents — the long, technical climb of hardware; the tethered, steep climb of software; and the guided, well-paced climb of applications — each of them dependent on the others and ideally converging at the peak. Nobody was sure what you'd find there.

The hobbyist revolution that Byte chronicled through the 1980s brought computers into everyday life, and our experience of computers today is largely defined by applications. Indeed, the revolution has come full circle so that networked computers have become what the mainframe once was — only now it's the cloud, and computers are hidden in the mist.

"The computer has become an appliance," said Jason Kridner, the developer of the BeagleBoard. "The machine loses relevance if it can't interact with the physical world, if it sits in the corner and just connects to the internet." Kridner remembers the computer that he had as a youth. "My mom took the floppy disks and put them in a safe, so I could hack that computer top to bottom." Like Eben Upton of Raspberry Pi, Kridner wants to bring that kind of computer back.

Kridner was an electronics hobbyist growing up, reading Forrest Mims. "Using a microcontroller to blink an LED would be the stupidest thing to do," he remarked. "I'd use a 555 timer." He started developing BeagleBone to satisfy his own goals and help out Texas Instruments as well. His target was Linux developers. "The goal was to put in their hands a platform that would allow them to do new things to advance Linux."

"I didn't know about the maker market, per se," said Kridner. "Yet when makers started picking up the board and doing crazy, fun things, the lights went off." At Maker Faire Detroit, near Kridner's home, there was a pick-and-place machine by Jeff McAlvay and a security device by Phil Polstra, each powered by BeagleBone. The OpenROV project, featured in Volume 34, also runs on BeagleBone.

In this issue, we chronicle a second hobbyist revolution that's starting small with new hardware — a growing number of credit card-sized microcontrollers and processors including Arduino, Raspberry Pi, and BeagleBone.

"What interests me is seeing technology connecting to everyday life, not just stuff in the cloud," Kridner said. "It's about taking away the mystery of computers and allowing people to build things out of electronics." Projects like ArduSat, an open source CubeSat satellite (see Volume 24, DIY Space), and the Earth-imaging satellites from Planet Labs demonstrate that it's possible to get above and beyond the cloud. ◢

Dale Dougherty is founder and CEO of Maker Media.

James Burke

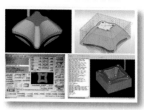

Siphon Logic, DTV Antennas, Arduino Sound Synthesis, and Japanese Toolboxes Galore.

» Nice article on siphons in the latest issue of MAKE ("Ctesibius and the Tantalus Cup," MAKE Volume 35, page 134) but way off on science. A siphon using water will not work in a vacuum. The example you gave is totally unfair. The example uses a special fluid with extra properties not considered in a classical siphon. Take a tube 30 feet long, closed at one end. Put the open end at the bottom of a reservoir of water. Draw a vacuum in the top and the water seems to be pulled up the tube, but by what force, cohesion? No. It is pushed up by the pressure of the atmosphere. Though for the "super" fluid used in the video, cohesion works for that case; it is not correct to suppose that atmospheric pressure is not working in other cases. Bad logic and bad science. Tsk! Tsk!   —Dr. Arthur G. Schmidt, Evanston, Ill.

MAKE TECHNICAL EDITOR SEAN RAGAN RESPONDS: Whoops! Our bad. The video Bill mentions does actually demonstrate a siphon that works under vacuum. However, it is not a water siphon; it's an ionic-liquid siphon. Ionic liquids have much stronger intermolecular attractions than other liquids, and it is likely this unusual property that makes it possible for them to siphon under vacuum. But we were wrong to infer, from that special case, that atmospheric pressure is not involved in the siphon effect with more familiar liquids like water. Thanks for writing and keeping us honest!

» Not everyone will take it the same way, but the "Water-to-Wine Cooler" project (Volume 34, page 46) could be construed as honoring (not disparaging) Jesus' first miracle, which itself had an element of humor in it (surprising a crowd).
    —Michael A. Covington, Ph.D., Athens, Ga.

» A couple years ago I made your DTV coat hanger antenna (vimeo.com/2931902) — my shirts are still on the floor of my closet — but never hooked it up until last night when PBS was supposed to air the remake of Hitchcock's *The Lady Vanishes*. It worked great for picking up our public TV station about 35 miles away. Wish I had done this long before. Thanks for publishing even though the jerks at PBS sired something else during their pledge drive.
    —Jock Ellis, Cumming, Ga.

» Thanks to MAKE and Jon Thompson for the "Advanced Arduino Sound Synthesis" article (Volume 35, page 80). Though I don't have much interest in sound synthesis, this was a little more in-depth than the typical Arduino article, making it more interesting. Please consider doing other more advanced Arduino articles.   —James Matthew, Sheffield, Vt.

» I'm a professor of mechanical engineering at the University of Colorado. I really enjoyed Jon Thompson's article on using the Arduino and advanced interrupt code to synthesize waves. I like it so much that I used the article as a lab in my System Dynamics (senior level) course. Actually, it's kind of funny, I've been developing new labs for this course all summer and was at a point where I didn't quite figure out the next lab idea for the following week and then MAKE 35 landed on my door! Just in time!
    —Shalom D. Ruben, Ph.D., Boulder, Colo.

EDITOR'S NOTE: Shalom shared the lab with us: makezine.com/go/lab.

Rob Nance

Richard White

Scott W. Vincent

Michael Levy

Greg Kent

EDITOR'S NOTE: Master woodworker Len Cullum's Japanese toolbox project (MAKE Volume 34, page 110) inspired a lot of readers, who shared their builds with us. Make your own at makezine.com/projects/make-34/japanese-toolbox.

» This was a great beginner's project. I didn't have a miter saw big enough to cut the 1×12, but it still worked out great. I also used red oak for my build. It cost a bit more and is heavier. Once I made the first box and did some reading and just studied the box, I was able to make the second one with my kids. The smaller one was made using only glue.
—Richard White, Bowie, Md.

» I really enjoyed this project and I'm very happy with my completed toolbox. Thank you Len for this wonderful project! Here's the information on mine: lungstruck.com/projects/japanese-toolbox. The only thing I did to make it a little more unique was chiseling my first name onto it in Japanese.
—Scott W. Vincent, Geneva, Ohio

» Really functional design. I made an 800×400×400mm toolbox for my van from a single sheet of 17mm ply. Total cost was $65 (Australian) and ~4 hours work. Much stronger than any of the sheet metal toolboxes you can buy for ~$200.
—Michael Levy, Wollongong, NSW, Australia

» I love it, my daughter loves it. Thank you MAKE! —Greg Kent, Kailua, Hawaii

AUTHOR LEN CULLUM RESPONDS: I can't tell you how this makes my day. Seeing that other people have been moved to make something because of an article I wrote is way more gratifying for me than the result of building it myself. Really. I am so honored that this is happening. I would happily shed my client work to get others to try woodworking/making for themselves. Thank you to everyone at MAKE for giving me the unexpected opportunity to experience this.

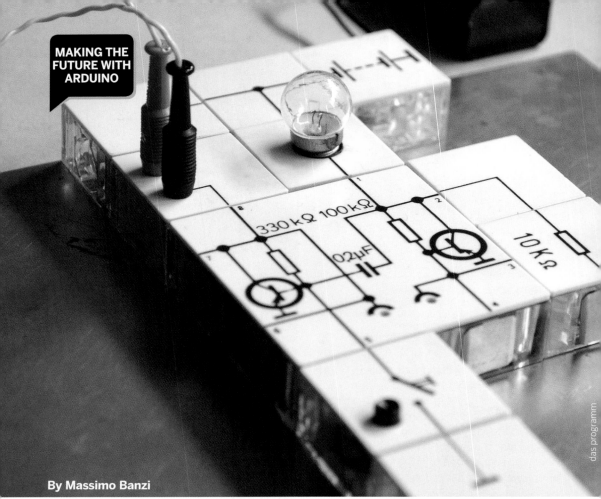

das programm

By Massimo Banzi

People Over MEGAHERTZ

When I was a kid I got into electronics because I started reading specialized magazines on the topic. At the same time it was hard for me to learn electronics from them because the content was not really beginner friendly and the projects were not very exciting. They were conceived more for people who were already into the technology and loved circuits than for explaining to newbies what circuits do and what you can do with them.

The way I really started learning electronics was when I received a kit as a present. It was called the Lectron System and was made by the German company Braun. It was composed of cubes you could snap together magnetically to build different circuits just by following some simple drawings and instructions. The cubes were transparent, so you could look inside to learn about the electronic parts.

The kit was a complete experience because it also had a book with great illustrations and simple explanations designed to look very appealing and make technology less scary through hands-on experiments. The original ad said: "Hey look, I just built a radio in two minutes" and it was actually true!

Designing the User Experience

The most interesting aspect of this kit was the ability to shorten the time between starting a project and the moment you get a positive result right "out-of-the-box." Playing with it and learning from it got me into electronics and sparked my interest in design.

The kit was indicative of a design style championed by one of the most important designers of the period: Dieter Rams. He worked for Braun in the '60s and '70s and created many iconic objects (including the packaging for Lectron) and inspired a lot of contemporary Californian design.

Dieter's way of looking at design was expressed in a broader sense: He came up with a list of design principles, and many of those principles reflected the relationship of people interacting with objects and space. I think this point is very important when designing technology: We must care about the people who are using it more than the technology itself.

When I got my first computer in the '80s, it was the moment when people could finally afford a computer without mortgaging their home. To use it I had to punch hexadecimal numbers on a keyboard, resulting in numbers displaying on the LCD display. It was an Amico2000 (Friend2000), and it was not what I'd define as "user friendly."

My Sinclair ZX81 Basic was a great improvement. It had only 1KB of RAM, but I could do a lot of stuff with it. It was really simple and could offer a whole experience. Even when I took it apart (a habit I've had since I was a kid), the circuit gave me a feeling of simplicity from just a few components you could assemble yourself.

The book that came with it — even if you happen to read it now — offers a good way to learn the basics of the programming language by moving forward progressively toward more complex concepts.

The Birth of Arduino

Fast forward to 2002. I was teaching at IDII Design School in Ivrea, Italy, the city where Olivetti was born and a lot of the Arduino boards are still made. The school was focused on interactive design, a specific branch of design that looks at how people work and connect with technology—the idea is to not only design the shape of something but also how people will engage with that object. This is very important because you can have a nice product with a terrible interface, and the result is a less-than-beautiful user experience.

The school's students usually don't have a background in technology. They don't know how to program or to do electronics, and we only gave them two to four weeks to create physical computing projects. At that time, the tools you'd find in the market were mostly designed for engineers, with a lot of options, lots of jumpers, and lots of connectors. Students found them too complex and couldn't figure them out properly. Looking at the way we worked with students taught us a lot, and Arduino came out of that work.

Optimizing User Experience

If you look at it, you realize Arduino boards are a mashup of open technologies wrapped up in a unified user experience. From the out-of-the-box experience we want to know how long it takes you to go from zero to something that works. This is very important because it creates positive reinforcement that you are on the right path. The longer that time is, the more people you lose in the process.

I think we are all on the edge of a new step in the Maker Movement, and some of you are surely working on the next big thing. Please keep at it, but keep in mind the overall experience. You can put in a processor that is 100MHz more than another one, but the way you interact with it makes a huge difference to people. It's more important to take care of the experience people have when they learn than to give them power they don't know what to do with. ◼

Massimo Banzi is co-founder of the Arduino project.

➕ Pick up your Arduino Robot today at the Maker Shed (makershed.com).

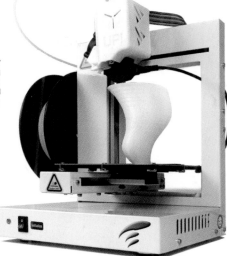
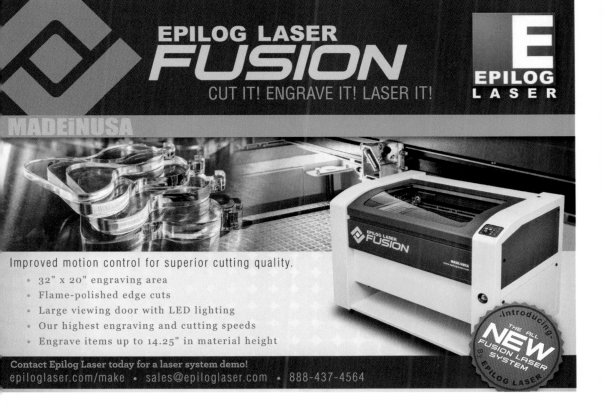

Experiments in 3D Printed Fashion

By Syuzi Pakhchyan

Dita Von Teese makes 3D printing look gorgeous in her custom nylon dress, designed by Michael Schmidt (not pictured) and Francis Bitonti (lower right).

Fashion and technology are the perfect pairing. Fashion's constant need for newness drives designers to continually seek novel materials and innovative ways of making garments. Emerging technologies provide designers with fresh inspiration and opportunities to innovate and create radically new garments that don't already exist.

Just a few seasons back, the laser cutter was fashion's muse as designers sent leathers down the runway, laser-cut in motifs of shattered glass, sporty mesh, and houndstooth. The laser cutter offers designers a new process to inexpensively fashion textiles with intricately detailed patterns, allowing them to transform leather into lace and acrylic into jewelry.

This year the 3D printer has taken center stage. Today 3D printing technology and fashion are a clumsy fit. The limited palate of available materials is too rigid, the resolution too crude, and cost too high. Yet this emerging technology has sparked fashion designers' imaginations because it radically changes the way garments are designed and made. Instead of hand-drawn sketches and two-dimensional hand-cut patterns, designers can digitally fashion three-dimensional computer models of their garments. This process provides them the freedom to create clothing and shape silhouettes that couldn't have been made in the past.

For Dutch designer Iris Van Herpen, 3D printed fashion and accessories have become her signature. One of the first designers to

Albert Sanchez (Dita), Bitonti Studio (rendering), Jeff Meltz8

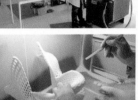

present laser-sintered fashion on the runway with resounding success, Van Herpen's experimental work is paving the way for 3D printed clothing. Her 2011 Paris debut collection *Capriole* presented a sublime dress that hung on the model like a fragile exoskeleton. Unlike previous seasons, her latest collection *Voltage* inches closer to more wearable forms.

With 3D printing, designers are gambling on innovations for new, flexible materials. Van Herpen's success and creative collaborations with companies like i.materialise have worked to pool resources and intelligence in, resulting, for instance, the development of "TPU 92A-1," a rubber-like material that is elastic, lightweight, and durable.

As Van Herpen pushes to develop new printable flexible materials, other designers are engineering structures to create flexible surfaces from solid nylon. One beautiful example is the 3D-printed bikini "N12," developed by Continuum Fashion in partnership with Shapeways. Using a clever circle patterning system connected with thin springs, Continuum Fashion's printed garments flex where needed to allow movement. The bikini can also be easily adjusted to any fit because the pattern is created using an algorithm.

Designer Ron Arad is also using 3D printing to produce glasses for eyewear brand PQ. The glasses are made from a single piece of material with hinges created by simple scores, allowing the frames to flex.

For Dita Von Teese's sensual dress, New York designer Michael Schmidt and architect Francis Bitonti created a fully articulated mesh design based on the Fibonacci sequence, digitally tailored for Von Teese's curves and 3D-printed in surprisingly flexible nylon by Shapeways.

The real impact of 3D printed fashion won't just be a new take on digital tailoring, but its potential to provide a sustainable means of manufacturing garments. Since 3D printing is an additive technology, it cuts out any waste in production. More importantly, garments can be made-to-fit and made-to-order locally, eliminating the need for excess stock and transportation of the product.

It's too early to proclaim that 3D printing will disrupt the fashion industry the way the internet disrupted the music and film industries. But as young emerging designers with untamable imaginations and an ethos for working in sustainable fashion continue to push the technology, 3D-printed clothing may just become a widespread reality. ◪

Syuzi Pakhchyan is a fashion technologist and author of the first DIY book on interactive fashion, Fashioning Technology (fashioningtech.com)

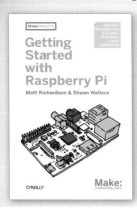
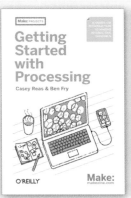

Captain Nemo's
Dream Machine

Becca Henry

By *Goli Mohammadi* **fivetoncrane.org**

Oakland, Calif.-based artist collective Five Ton Crane is in the business of bringing fiction into reality through their incredibly detailed large-scale works. Their latest creation, a collaboration with **Christopher Bently** and lead artist **Sean Orlando**, is the land-based *Nautilus Submarine* art car, a nod to *20,000 Leagues Under the Sea.*

Gregory Hayes

Built upon a 2005 Eagle TT8 diesel AWD airport tow tractor, the piece weighs 11,000 pounds, is 25' long, 8' wide, and 11' 6" tall, with a top speed of 13mph. Features include a harpoon gun water cannon that can shoot 13 gallons per minute, hydraulic drive controls, air conditioning, a night-vision periscope, a quaint library and map room, a killer sound system, working brass apertures, a programmable lighting system, and, of course, a specimen lab. No detail was overlooked inside or out, creating a fully immersive experience upon entry.

Gunther Kirsch

The Nautilus stole the show at Maker Faire Bay Area 2013, even serving as a stage for Adam Savage to give his popular talk to a packed crowd. Captain Nemo would definitely be jealous. ◪

Karen Kuehn

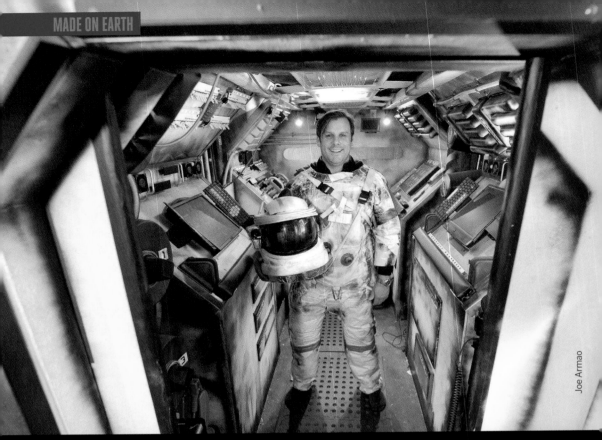

Joe Armao

SPACE STATION
in a SHED

By *Laura Cochrane* mkoutlier.com

To the casual passerby, **Christopher Jacobs'** backyard shed appears nondescript. But beneath the worn white exterior exist hundreds of switches, blinking lights, computer screens running animations, and a green-screen viewport — all part of a space station movie set Jacobs built himself.

The project began when the West Footscray, Australia resident wrote a sci-fi film script, called *MK Outlier*. He then constructed this space station set in his backyard, with the help of his friend **Harry Hortis**. Jacobs even crafted a space suit costume for the film, with help

tape." In fact, the film was one big DIY side project for everyone involved, with the entire cast and crew working on off-hour evenings and weekends to complete the shooting. The shed space-station frame was built by Hortis, and then Jacobs took over, crafting the look and feel of a space station using found objects. "My girlfriend's dad's old pool pump is in there somewhere," he says.

The most challenging part of the build was working in a tin shed under extreme weather conditions. "In summer it was an oven, and in

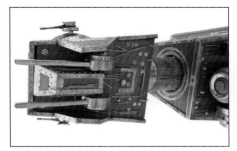

Caitlin McDivett

The Imperial
Drinker

By *Goli Mohammadi* **colinrhino.blogspot.com**

Vancouver, British Columbia artist and "woodbutcher" **Colin Johnson** spent roughly 600 hours handcrafting this brilliant pony-sized AT-AT liquor cabinet. Naturally a *Star Wars* fan, Johnson was inspired by the aesthetic of the Empire's machinery. *The Emperor's Cabinet*, made from high-density plywood, mahogany veneer, solid brass trim, and glass, stormed through the Vancouver Mini Maker Faire in June. ⬈

Twist OF Fate

By *Goli Mohammadi* mikerossart.net

Los Angeles-based artist **Mike Ross** is fascinated with power in all its forms, be it physical, political, economic, or the power of humanity to make its own world. In 2007, Ross' vision manifested in the awe-inspiring form of a 50'-tall, 25-ton sculpture constructed from two discarded tanker trucks, one precariously perched atop the other in an unnatural scorpion-like twist.

Aptly titled *Big Rig Jig*, the piece took a dedicated crew of seven full-time folks and numerous community volunteers three months to build. The finished piece is not only a mind bender to view, but serves as an architectural space, inviting visitors to enter the lower truck, climb up the tankers, and get a birds-eye view from the viewing platform 42' up on the rear axles of the upper truck.

He's currently working on a sculpture made from two decommissioned U.S. Navy Skyhawk jets. Let the imagination begin! ☑

Ryan Jesena

Hot
Bots

By *Goli Mohammadi*   **gileswalker.org**

British artist **Giles Walker** has been making provocative art robots and kinetic sculptures for more than 20 years. His *Peepshow* installation features two metal and plastic life-sized pole dancers with CCTV heads who gyrate as the bullhorn-headed DJ drops beats between them. Built purely out of scrap using windshield motors and Wizard 3 boards, *Peepshow* is Walker's reaction to the increase in the number of surveillance cameras in England, choosing pole dancers as a play on voyeurism and its relationship to power. Walker adds, "I also was interested in the challenge of whether I could make a pile of old scrap, sitting in the middle of my workshop, into something sexy!" ◪

Giles Walker

HIGH FLYING

Hope

Written by Jon Kalish

PETER LYNN'S VISION FOR THE FUTURE HANGS FROM A STRING.

New Zealand kite maker Peter Lynn has a 30-acre spread in Ashburton, about an hour south of Christchurch. Walk with the 66-year-old engineer from the house he lives in with his wife Elwyn, past his workshop with a sign that says "The Engineerium," continue up the dirt road to the kite-making factory, and it's hard not to notice that there are other, er, enterprises on the land. Tenants include a kitchen builder, a junkman, a recycling center, and "panel beaters" — what is referred to in America as auto collision repair. In such an environment, therefore, it's no shock to come upon a 1918 Bessemer oil well pumping unit. Lynn seems to take pleasure in the fact that it's not green. "We thought we needed one in our backyard," he wisecracks.

Lynn brings me into an old barn that was used to store timber for his father's woodworking business. Now it's home to "the archive" of his 40 years in the kite business. "There are some crazy ideas

WHALE WATCHING: Lynn's 52'-long blue whale kite — also available in 26' and life-sized 98' lengths — flies over Lake Clearwater in New Zealand in June 2013.

here that you wouldn't want to see," he jokes. Such as the "Jesus feet." Nearly five feet long, they enabled humans pulled by a kite to glide along water. It was one of Lynn's early efforts in "kite boating." "You strap them on your feet and off you go," Lynn explains. "And it worked, too, but it was hard to steer."

Elsewhere in the barn is a three-ton

Peter Lynn

Jon Kalish

Campbell split-frame portable engine circa 1902, which Lynn says was extracted over a period of nine years from a swamp where it had been abandoned in 1931. "It took me about the same amount of time to put it back together and replace a few parts. And now I take it out to the old engine shows and run it," he says of the oil engine. "It's very rare. I think there may be another one in the world but there probably aren't two others."

Lynn's face lights up when he shows off a Stirling engine he built. It's stationed inside a 26-foot-long canoe-like vessel, a replica of a 19th century Great Lakes fantail launch. "I heat it up and away we go," Lynn says of the launch, which is named Piwakawaka. On the floor near Piwakawaka is a small yellow craft that is part of Lynn's continuing efforts to perfect kite sailing.

Lynn believes that kites can do the job of moving

1. Brian Holgate pilots the Peter Lynn Speed Buggy pulled by a nearly 9' Lynn Vapor kite at Ivanpah Dry Lake Bed in California during the 2012 series of runs, breaking both the kite buggy speed record and the overall kite speed record.

2. Lynn's 78'-long cuttlefish kite flies over the Adelaide Kite Festival.

3. One of the world's largest kites, owned by the Kuwaiti kite team, is Lynn's 13,455-square-foot New Mega Ray.

4. Hand-sewing giant kites in the workshop.

5. Humans standing in the New Mega Ray's mouth show its large scale.

"It was a toss-up whether I'd be cut in half by the kite lines before drowning or the other way around."

a boat through water better than sails do. "It hasn't happened yet," Lynn says, "but I still think that." He's been experimenting with traction kites to pull people in various vehicles and vessels since he was a kid. The results have been both triumphant and disastrous.

As a boy he used kites to pull carts and bicycles up and down the roads of Ashburton. Once, while flying kites at a school playground, young Peter Lynn rode his bike full speed into a kite line and cut his throat. "I had a good go at cutting m'head off," he says in his thick New Zealand accent.

In 1987, he began devoting much of his professional energy to perfecting traction kites for power boats, buggies, boards, and sleds. In late 1987 or early 1988, Lynn was on Lake Clearwater about 45 miles northwest of his home

in Ashburton, experimenting with a catamaran and a rerigged skydiving parafoil. There was as much as 1,000 pounds of pull on the kite's line when Lynn went blasting off into the lake. The boat flipped over and he was dragged underwater, trapped between the kite lines and the catamaran's hull. As he was pulled backward at 15 knots, a big rooster tail of water marked the vessel's progress across the lake.

"It was a toss-up whether I'd be cut in half [by the kite lines] before drowning or the other way around," Lynn recalls. His frantic efforts to break free from the nylon lines eventually got him to the point where he could take occasional breaths, and soon Lynn managed to twist free and begin the long swim back to shore. He estimates traveling close to a mile before escaping. The boat continued up the lake, through a swamp and then bounced along the countryside for another mile before getting stuck on a rocky hill with the kite still aloft.

But despite such mishaps, Lynn's contributions to the development of power kiting are major. In

1990, when it got too cold to sail on the lake, Lynn converted a boat that had three skis into a buggy with three wheels and rode it around on land. "Well, this works very well," he said to himself, and a few weeks later he was making kite buggies that are steered with the feet. They became a hit in Europe, despite what Lynn describes as an effort by Europe's "kite mafia" to squelch his foray into the market. Lynn estimates there are more than 10,000 of his buggies tooling around on the sand and other environs. Lynn also worked on a kite sled, and one he designed was used by two Australians to make a 435-mile journey across Greenland in 2006.

Lynn will tell you that engineering is a big part of kite making, and he soon realized that a better kite was needed for his buggies, so he developed a two-line, steerable kite. Improvements in the materials used to make traction kites have had a profound impact on progress in the design. The traction kites designed by Lynn are made of synthetic fabrics,

6

6. Lynn pilots his ice buggy across frozen Lake Clearwater, drawn by his 26' adaptive-profile foil kite prototype.

7. The Speed Buggy, with its low rolling resistance and streamlining, is efficient at multiplying wind speed and can reach over 75mph in a 31mph wind.

8. Lynn's hand-built Stirling engine and his fantail launch.

9. Lynn's rare three-ton Campbell split-frame portable engine, circa 1902.

7

8

John Kalish

including Spectra, which has immense strength.

Ask Lynn to explain the physics at play here, and he starts referring to stagnation points, air velocity, and Bernoulli's theorem. But the bottom line is that the traction kites Lynn has designed make buggies fly like a bat out of hell. To wit: The current world record for kite-powered travel was set on March 6, 2012, in the small California desert town of Ivanpah, where winds were gusting at 55mph. Brian Holgate went 84mph in a speed buggy designed by Craig Hansen, Gavin Mulvay, and Lynn. The vehicle was powered by a Vapor 2.7 traction kite designed by Lynn. The kite's two lines are connected to a strutter bar that runs between the driver's bellybutton and hips. Lynn says he expects kite buggies to reach a speed of close to 100mph some day. But Holgate, who works in TV production in Las Vegas, is gunning for even higher speeds. It may be pie in the sky but Holgate thinks a kite buggy will someday surpass the 126mph record set in 2009 by the British team behind Ecotricity Greenbird, a land yacht with a solid sail that holds bragging rights for the fastest wind-powered vehicle on land.

At the moment, one of Lynn's pieces is in the Guinness World Records as the world's largest kite. The 11,000-square-foot build in the shape of a Kuwaiti flagtook the title in 2005, beating out the previous record holder, a 6,800-square-foot monster Lynn made called Megabite, which bears a distinct resemblance to an extinct marine arthropod. Lynn's New Mega Ray is even larger, but not officially listed in the record book.

Lynn has also pioneered the creation of soft kites, which are inflated and have no frames. If you've never seen them soaring in the sky, they look like the huge balloons in Macy's Thanksgiving Day Parade, in New York City. But instead of resembling cartoon characters, these display kites are often made in the shape of sea creatures. Lynn has rock star status at the kite festivals where such kites are flown. He travels frequently to fly his creations around the world, including at the premier event held in Berck sur Mer on the French seacoast. Close to a million spectators turn out for it over a nine-day period every April.

After more than 40 years in the kite-making business, Lynn seems to have accepted his life-long obsession with these airborne creations. He once struggled with the decision whether to go into business manufacturing combined heat and power units but now says he's "quite glad" he didn't. Ditto for a career making portable sawmills. "Doing what I'm doing now," he declares, "isn't boring." ◪

9

➕ www.peterlynnkites.com

Jon Kalish is a Manhattan-based radio reporter and podcast producer. For links to radio docs, podcasts, and stories on NPR, visit Kalish Labs at jonkalish.tumblr.com.

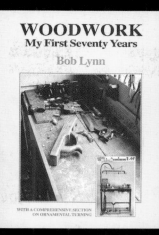

WOODWORK
My First Seventy Years
Bob Lynn

WITH A COMPREHENSIVE SECTION
ON ORNAMENTAL TURNING

Meet the Lynns:
An Energetic Clan

Making seems to run in the veins of the Lynn family. Peter's father Bob Lynn, who passed away at the age of 97 in early 2012, was a carpenter who built houses and also did woodwork in the New Zealand parliament building, using native timbers kohekohe and mangeao. He established the Museum of Woodwork, Ornamental Turning and Tools in Ashburton and wrote a book titled *Woodwork: My First Seventy Years*.

Peter and his two sons, Pete and Robert, earned engineering degrees at the University of Canterbury in Christchurch. Both his sons have left New Zealand, and 39-year-old Pete Lynn is listed as Chief Seamstress Officer of Otherlab, a San Francisco engineering firm founded by MAKE columnist Saul Griffith. Pete met Griffith in the kite-surfing scene.

"I basically develop weird stuff," Pete says, before elaborating that his duties at the lab involve inflatable robotics (featured in MAKE Volume 34) and alternative energy. One of his projects at Otherlab is focused on moving small mirrors around to optimize the harnessing of solar energy for electricity production.

Pete's older brother, Robert (age 41), lives in England and is also devoting his engineering talents to an energy project — a heat pump (a device used to pull heat out of the air or ground for the purpose of heating or cooling a building). "It involves some thermodynamic trickery," he explains. Robert had previously been employed developing engines for expensive European sports cars. (Robert built his first engine at the age of 13 with his father.) His interest in internal combustion engines, he jokes, is "a family disease."

Tiger GATE

Written, photographed, and illustrated by Tim Hunkin

In the past, the zoo had always used horizontal sliding gates. But the relentless rise in health and safety standards had led to the design of an elaborate new warren of cages for the tigers. The simple horizontal gates would not fit in five of the locations and these gates would have to slide vertically. Initially I couldn't understand why this was difficult; I only later realized the problems involved.

The architect and keepers had visited the Vienna Zoo, which has sophisticated electronic vertical gates. I suspect the London Zoo couldn't afford them and was looking for a cheaper solution when they contacted me. My initial instinct was that electronics were a bad idea anyway, because if one failed the zoo's maintenance team would have to call out specialists. If the gates were entirely mechanical, the maintenance team would be more likely to be able to mend the gates themselves. This is KISS technology — "keep it stupidly simple."

My mechanism design was indeed simple, but it was more important that it worked. For example, I used plain nylon strips for the slides rather than rollers, which could potentially jam. I used polyester rope instead of steel because I've had trouble with strands breaking. It's the same sort of failsafe approach that is used in nuclear engineering. I presented my drawing and was delighted that they accepted it.

With the design worked out, you might think everything else would be straightforward. In practice, it was only the beginning of the journey. The first problem was trying to fabricate the frame accurately. This was important because the channels had to be completely parallel for the gates to slide smoothly. Every weld distorts the frame a bit so it's a real skill working out where to clamp things and what order to weld everything. Later on I realized it was better if the gates were a really loose fit, so this fussing was mostly unnecessary.

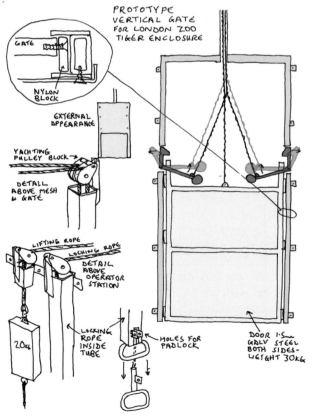

PROTOTYPE
VERTICAL GATE
FOR LONDON ZOO
TIGER ENCLOSURE

GATE

NYLON
BLOCK

EXTERNAL
APPEARANCE

YACHTING
PULLEY BLOCK →

DETAIL
ABOVE MESH
& GATE

LIFTING ROPE

LOCKING ROPE

DETAIL
ABOVE
OPERATOR
STATION

20kg

LOCKING
ROPE
INSIDE
TUBE

HOLES FOR
PADLOCK

DOOR 1·5m
GALV STEEL
BOTH SIDES-
WEIGHT 30kg

NOT CATCHING A TIGER BY THE TAIL

The gates at the Vienna Zoo were solid to avoid finger (or rather, claw) traps. With 1.5mm-thick steel plates on both sides of my gate, I was shocked to find it weighed over 30kg. It was certainly too heavy to lift comfortably by rope. The whole thing looked scarily like a guillotine and Graham Norgate, who was helping me, pointed out that I needed the locking device to keep the gate open as well as to keep it closed. The keepers had been nervous about vertical gates, not because the tiger might escape, but the gate might fall on it — and I really didn't want to be responsible for crushing a tiger's tail.

I added counterweights, but then the gate wouldn't close when the rope was released. So I added an extra pulley. This doubled the mechanical advantage. The rope is pulled twice the distance but with half the effort. On day three, we tested the prototype. It just about worked but the lock wasn't easy to use and the gate was still stiff sliding. After another day it was ready to take to the zoo

EXTRA
PULLEY

to show the keepers.

At the zoo, a Polish builder named Przemek Tuliszka helped me assemble it for the trial. He thought the keepers were crazy. "We are doubling the thickness of walls but what does a tiger weigh? Maybe 120kg — that's much less than a car, it's nothing."

CATPROOF? YES! BEARPROOF? NOT SO MUCH

Eventually lots of keepers and officials arrived. After a short demonstration they tried the gate themselves. A fierce Scottish keeper kicked it as hard as he could, not leaving even the slightest mark — very satisfying. Without unlocking the gate, he then pulled on the lifting rope as hard as he could and managed to move the locking bars a bit (though not nearly enough to release the gate). His verdict was that the design was just OK for a big cat, but he wouldn't have passed it for a bear. Cats are basically lazy and give up quickly but bears can persevere for ages.

"I REALLY DIDN'T WANT TO BE RESPONSIBLE FOR CRUSHING A TIGER'S TAIL."

Despite being passed for strength, the gate failed its trial by keepers for a completely different reason. I had imagined tiger keepers would be strong men, but at the London Zoo the mammal keepers work as a team and at least two of them are small women. They found my gate too heavy to lift, even though we had added the extra pulley to make it easier. (The attitude to lifting things has changed radically in my lifetime. Coal used to be carried in 100kg sacks and cement came in 50kg bags. The maximum sack weight was then reduced to 25kg and now workers can refuse any load if there is a possibility of back injury.)

We couldn't add more counterweights or the gate would not have enough weight to close. The only possibility was to decrease the friction. I assumed the main source of friction was the slides, but we quickly found the problem was the pulleys. These were standard yachting pulleys with plain bearings. Fancy yachting pulleys with ball bearings are available, but the bearings are open. This is fine at sea but not suitable for a cage where dirt and straw could get in. So we made our own pulleys with sealed ball bearings. Satisfyingly, these almost halved the effort needed to lift the gate (from 15kg to 8kg).

CAT'S CRADLE: Prototyping a mechanism like the tiger gate is a combination of the usual R&D and collaboration with the zookeepers, who are experts in animal behavior. Many pulleys, counterweights, and trials by zookeepers led to the final design.

"ONCE FINALLY ON-SITE, EVERYTHING WAS CHAOS."

Two keepers, including their smallest, had agreed to come to my workshop for the second trial. She passed it as easy enough to use, and then suggested several sensible modifications, particularly marking the two positions of the locking rope "locked and unlocked."

HOPES AND SCHEMES
My original plan had been only to make the prototype and pass the design on to a company to manufacture and install the final gates. However, as the design evolved, I had become increasingly anxious that this could go horribly wrong and decided to install them myself.

To prepare for manufacture, I drew all the parts in CAD, but the company I was working with, Eastern Hardware, said they would prefer to have the prototype in their yard and work from that. This is also how I prefer to work, but it does require the fabricator (in this case, a fellow named Brian Orvis) to be a "schemer" who prefers problem solving to rigidly following drawings.

I immediately trusted Brian and was lucky because he did a brilliant job. We spent two days working together assembling everything after the parts had been galvanized. Like Przemek, he also thought the design was a bit "over the top." He had just watched a TV program about a New Zealand man who kept tigers as pets. He claimed they were perfectly docile if never fed raw meat (always cooked meat) and given a

daily bath to stop hormones building up.

The installation itself was not straightforward. I had to pack everything I might possibly need because it can take hours going off-site to buy anything. In the past, when I often worked away from home, my van was full of useful stuff, but now I'm out of the habit and my van is empty. Also the tools posed a challenge because, unlike most U.K. tools, which are 230V, building sites require 110V tools, and I don't have any. I rented some 110V tools and packed my generator in case we also needed one of my 230V tools.

I'm not used to working on building sites. In advance it was all very formal: the architect's drawings, the paperwork for the risk analysis and method statement, the obligatory safety clothing, and the induction lecture. However once finally on-site, everything was chaos, though fortunately a benign chaos. It was midwinter, and after a wet autumn the site was alternately deep mud or covered in snow and ice.

MAKING MY MARK
We had been there less than an hour when we disgraced ourselves. A mover had driven our gates close to their location. Here there was a handy pallet truck so we transferred everything and pulled it the rest of the way. A very bad move. The concrete under our feet had only been poured the day before so the wheels left an indelible trail. I had to go and confess. Lots

FROM FAB TO INSTALL: Fabricator Brian Orvis (far left) manufactured the gate using the prototype instead of CAD drawings, the old fashioned way. The final installation and testing took several days due to unforeseen circumstances. In the end, though, the build was a success and the London Zoo tigers have shiny new gates.

of officials arrived and spent ages looking at the damage. I was given a strong reprimand, so for the rest of the installation I remained terrified what would happen next.

In fact, the next shock was the cage dimensions. Although the architect's drawings looked so precise, in reality the cages were 50mm lower and a beam the ropes had to pass under was 100mm lower. The team installing the cages was busy grinding bits off everything to make them fit, so we had to follow their example. It feels wrong to drill, cut, and weld steel that has been galvanized (dipped in molten zinc) to protect it against rust. Fortunately the galvanic protection works like a battery so oxygen attacks the zinc in preference to the steel, and this protection extends beyond the zinc to any exposed steel nearby.

It was a slow process though, and after three days (the time I'd estimated for installing the gates), we still weren't nearly finished. We returned the following week, having welded up some extra parts to make things fit. After another two days on-site, struggling through intermittent snow and rain, all the gates were finally installed and working.

The keepers then had to inspect and pass the work. Even though they'd seen the prototype, they expressed surprise that the gates were solid plate. Mesh would have been better so they could see the tiger on the other side. If we had used mesh instead of plate, the gates would have been so much lighter we would never have had to add pulleys and reduce friction. How unbelievably irritating!

So despite my initial enthusiasm, the job wasn't remotely glamorous. With the winter weather, the installation was quite miserable at times. And I never even saw a tiger until a month after I'd finished.

Still, I am proud of the gates. Some people do extreme sports, but my thing is extreme making. I was not the obvious person for the job — I normally make arty things. In the past it would have been easy for the zoo to find a company to design and make the gates.

There used to be thousands of small engineering firms in the U.K. But today, the engineering firms that remain have "modernized" and now work more formally. Even if willing to take on a small job like this, they would have charged a lot more money and been less flexible when faced with problems like the wrong size cages. The last generation of traditional engineers like Brian Orvis are retiring, so perhaps the future for jobs like this will lie with me and other makers. People who do it because they love practical problem solving and making things. ◪

Tim Hunkin trained as an engineer but became a cartoonist for a U.K. Sunday newspaper. He next made *The Secret Life of Machines* TV series and now runs an arcade of homemade coin-operated machines in Southwold, England.

KICK

FROM MAKER TO MANUFACTURER: RETHINKING HO

Only a short time ago, it would've been outlandish to think enough people would be willing to buy one of our weirdo projects to make it worthwhile for us to manufacture it. Today, thanks to crowdfunding, this scenario happens to makers all the time. It happened to my partner and me at our two-person design company, CW&T (cwandt.com), when we designed a pen and used Kickstarter to pre-sell 50 of them. When our campaign ended we had over 5,500 orders! Surpassing our funding goal by such an extent transformed our project from something that is *made* into something that is *manufactured*.

After spending a year manufacturing 6,000 pens, one of our next projects was a kickscooter. In designing it, we thought a lot about how the decisions we made would affect its manufacturability. We didn't have money to invest in tooling for custom parts. What if things didn't work out? What if we got bored with the project? We'd end up throwing it all away. So we designed the scooter using as many off-the-shelf components as possible, including stock 1" aluminum tubing and longboard and inline skate parts. By using materials that were lying around our studio, we had a prototype up and running in a couple of days.

HOLDING IT ALL TOGETHER

As we improved the design, we incorporated a few custom parts that required CNC

scooter starter

Written by
Taylor Levy

'E SCALE THE STUFF WE MAKE.

CW&T

machining to join the kickscooter components together, making the scooter more durable, easier to manufacture, and simple to assemble. The rest of it was made using tubing that the buyer can cut to size, an off-the-shelf longboard truck, and any set of longboard wheels.

Throughout this process, there were phases where we were so excited that we were tempted to try and sell tens of thousands of scooters as soon as we could. We've all had that feeling before. But this time we decided not to get too far ahead of ourselves. We wanted to find out if it was possible to manufacture and sell just 50 scooters in a way that was sustainable.

Since we designed the scooter to avoid high up-front costs associated with custom molds and tools, manufacturing 50 was actually possible because we were able to find a manufacturer with CNC machining capabilities who would work with us. Before CNC machining became commonplace in small production factories, it would have been nearly impossible to find a manufacturer who would even consider working with us at such a low volume.

In those days, the start-up cost for manufacturing was prohibitively high because most things were made by hand or using manually operated or mechanically programmed machines. To get up and

1. Early sketches of the minimal kickscooter, designed to be built with standard aluminum tube stock and off-the-shelf skate and inline skate components.

2. Testing out the first prototype with a clamp handle.

3. A later, machined prototype, with a handmade version on the floor nearby.

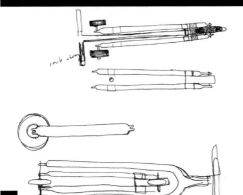

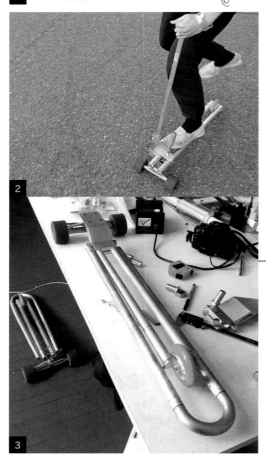

running, engineers first needed to design and build sophisticated mechanisms, jigs, and tools in order to make machine operations as efficient as possible while maintaining quality throughout production.

THE HIGH COST OF GETTING STARTED

Today, starting up is still the most costly part of manufacturing. As CNC machines become more advanced and ubiquitous at factories, getting a project up and running is much easier. But it's still far from trivial.

Engineers at factories now spend their time programming spindle speeds, tool paths, and changes on multiple axes of operation. These are among the many variables that are honed to the utmost precision. Every little detail matters when you're trying to program a machine to do work at its optimal capacity. It's nothing like the clean-cut computer programming we're familiar with. It's physical, messy, and dangerous if things go wrong.

Even with the most sophisticated machines, you still have to tune in with all of your senses. Do you know what a spindle at 4,000rpm running at 30"/min cutting 6061 aluminum with a three-flute bit on a Haas machining center from 2009 sounds like? Great machinists do. Getting this all right requires an incredible amount of knowledge and experience. It's a different type of work than it was in the past, but it's nowhere near as simple as pressing a few buttons.

That being said, every day machines

> "TODAY, STARTING UP IS STILL THE MOST COSTLY PART OF MANUFACTURING."

become more automated, versatile, and easier to program. As this happens, start-up costs will continue to fall and it will become more sustainable for manufacturers to take on smaller projects, like a run of parts to make our 50 scooters.

There's lots of valuable information to be gained in manufacturing 50 of something before jumping into a run of thousands of units. Sure, it doesn't sound as exciting, but the longer you can hold onto your project at a scale that you can manage, the better. No one cares as much about your project as you do. And the bigger it gets, the more you'll have to let go.

If you're used to being really hands-on, manufacturing can be a pretty jarring, uncomfortable experience — especially when you receive a massive shipment only to discover that everything needs to be trashed. Who knows what went wrong or why some detail was overlooked? When things go wrong, and with manufacturing they always do, it can leave you feeling pretty powerless.

THINKING SMALL

Compared to our last project where we jumped straight into manufacturing 6,000 pens, making 50 scooters felt like a much more sane, manageable way to see our next project come to life. We were using and learning the capabilities of some of the newest manufacturing technologies. We could stay involved in the process; 50 units was a low enough number that we could assemble scooters ourselves or with others as kits. And most importantly, we could begin to uncover some of the problems that would inevitably emerge if we ever decided to scale in the future.

The choices we made for the scooter project are just some ways to rethink the transition from maker to manufacturer. But there are still many other possibilities left to explore. Obviously one of the most transformative technologies in this domain is 3D printing, where the barrier for entry to start a production run is virtually negligible. If 3D-printed metal were as structurally integral and inexpensive as CNC machining a solid block, that is definitely how we would have made our scooters. To boot, we could've gotten away with manufacturing one or two or five units, instead of 50. How's that for

> "IF 3D-PRINTED METAL WERE AS STRUCTURALLY INTEGRAL AND INEXPENSIVE AS CNC MACHINING A SOLID BLOCK, THAT IS DEFINITELY HOW WE WOULD HAVE MADE OUR SCOOTERS."

different scales of manufacturing? We're not there yet for metal, but we will be soon.

Another approach to manufacturing is to build your own mini-factory. This might sound crazy, but it's really not that far-fetched. Some of the most inspiring work on this subject has come from a class called "The Low-Tech Factory: Experiments in Small Batch Manufacturing," taught by designers Chris Kabel and Tomás Král at ECAL University of Art in Switzerland. Some student projects included a rocking chair that knits hats, a multi-step workbench for producing a stamped light fixture, and a contraption that pops and salts one single popcorn kernel at a time. These projects explore a context where the manufacturing process, the product, and the makers are intimately bound. The factory becomes a playful and personal space that seems alarmingly welcoming, efficient, and sustainable.

With new projects in the pipeline, we should all be wondering, pining for, and experimenting with new ways to scale our making. We no longer have to shoot straight for the old world of mass manufacturing, and we certainly don't have to make everything with our own bare hands.

The scooter project is just one example of the many ways to evolve the culture of making. Slowly but surely, we will transform mass manufacturing as we've known it into models that are more intimate, thoughtful, and sustainable. This is now part of the creative process. It's up to makers, one project at a time, to redefine manufacturing. ◪

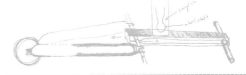

Taylor Levy (taylorlevy.com) is one half of CW&T (cwandt.com), an art and design studio based in Brooklyn, N.Y. Inspired by technology, her work simplifies complex or opaque systems by breaking them apart, exposing their inner workings, and reorganizing them into self-explanatory structures.

Interview by Goli Mohammadi

Meet LUCAS WEAKLEY

The creator of our Maker Hangar video series,

Seventeen-year-old Floridian Lucas Weakley has been an R/C aircraft enthusiast for the past eight years. Following his passion, he's acquired a wealth of knowledge and has started designing and building his own planes. In true maker spirit, he also shares his knowledge on his biweekly YouTube show called BusyBee TV, which features reviews, how-tos, and all things R/C aircraft. Lucas recently made a 15-part video series for MAKE called Maker Hangar, in which he walks us through the step-by-step build of a robust plane named the Maker Trainer.

How and when did you get interested in R/C aircraft?
I first got into remote control when my parents bought me a very large gas R/C trainer for my ninth birthday. I took lessons with it at our local R/C flying club.

And then you started sharing your knowledge through your videos—what inspired that?
A few years later, when I wanted to learn what all the parts were and how they worked, there was no clear source that I could go to. I would watch a YouTube video here and there and read a couple of forum posts. After several months, I finally had a basic understanding of all the parts and how they worked together. Then I thought it would be nice to have a one-stop place to learn everything. I had some experience with making my own videos for my YouTube channel, so then Maker Hangar was born.

How has being involved in Boy Scouts influenced you and your projects?
Being a Boy Scout has made me a better leader. It's given me the discipline that helps me do the right thing. Also, managing my Eagle Scout project prepared me to be able to handle these larger tasks.

When did you first start reading MAKE?
About nine issues in. I discovered the podcast first with the weekend projects by Kipkay and was hooked. I subscribed to MAKE right after and bought most of the other issues to fill in the gaps. I attended the 2010 Maker Faire Bay Area and have been an active member ever since.

What were some of the highlights of that first Maker Faire for you?
Maker Faire was an experience of a lifetime. I met so many cool people there, including all the people from the MAKE video podcasts. My dad and I had so much fun there, and having booster passes was worth every penny to come a day earlier and meet all the people and exhibits before the fairgrounds were mobbed. Meeting all the people who made the MAKE videos I love to watch inspired me to make videos and share my knowledge. If it wasn't for that, I probably wouldn't be here doing Maker Hangar today.

Check out all 15 Maker Hangar videos at
makezine.com/makerhangar

How has the Maker Movement affected and influenced you?

The Maker Movement has allowed me to learn so much as I was growing up. I learned so many of my skills and traits through MAKE, such as soldering, how to effectively use hand tools, how to program, and much more. Being a part of the Maker Movement has really helped me to become what I am today. To anyone with the slightest interest in making things, get involved — you never know where it'll take you.

What branch of engineering are you planning on pursuing?

I love aviation and all things airplanes. But I want to be a mechanical engineer because I love the process of designing and fabricating physical components. I would still work on airplanes and possibly earn a minor in aerospace engineering as well. I just want to keep my options for the future open and having a degree in mechanical engineering allows me to do that.

You own a 3D printer and work with CNC mills and laser cutters. How do they affect your builds?

I think having access to some kind of 3D printer or CNC machine is almost mandatory for someone wanting to design and make parts. The ability to design something and then an hour or so later have a perfect copy of that thing in your hand is unequaled by any other process. I'm able to print out my designs, make sure they're correct, and make changes to them all within a couple of hours. This allows me to rapidly prototype and design entire projects very quickly. My 3D printer is my most used tool, and I couldn't live without it.

What other tools do you swear by?

Hot glue gun and X-Acto knife.

Any advice for young makers who are just getting started with making?

Find what you love to do. It doesn't matter if it's R/C, Arduino, 3D printing, electronics, anything. Try them all, and see which one interests you the most. Then learn it, live it, and perfect it. If you put all of your ability in learning about what you love, first you should be having fun, and second you'll become a source who other people come to, and hopefully you'll contribute by making more innovations for others to learn from. You could also turn your passion into a career by offering a service that others need and you can supply. Turning your hobby into a paying job is something that everyone wishes they could do, and if you become the best at whatever you love, you can make that a reality. ◪

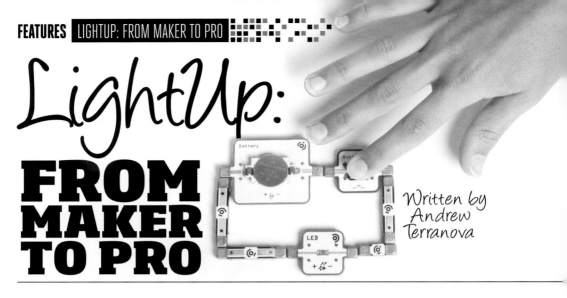

LightUp:
FROM MAKER TO PRO

Written by
Andrew
Terranova

HOW TWO CALIFORNIA KIDS WITH A BRIGHT IDEA ARE WORKING TO MAKE IT HAPPEN.

LightUp founders Josh Chan and Tarun Pondicherry are at a turning point. They are LightUp's only two employees, so far, but they're both working full-time now. They don't have an office, yet, and are punching the clock at home or in local coffee shops. That plus a website where folks can place pre-orders is about all there is to the infrastructure of LightUp, Inc. For now.

Their product is a modular electronic construction kit that uses magnets to snap together for easy, kid-friendly assembly. It's supplemented with a smartphone app that lets users see how electricity flows through their circuits and provides help through an interactive tutor.

It's supposed to ship in a matter of weeks.

THE BEGINNING

Pondicherry has been passionate about getting others involved in electronics since high school, when he taught basic circuits and robotics to middle-school kids.

"I kept seeing that it wasn't really the concepts that were hard," he recalls. "It was the platform that people eventually had to use — the breadboard."

Pondicherry and Chan met through a Stanford class called "Beyond Bits and Atoms" that focused on designing educational toys and interfaces. The final project was an assignment to design a "constructionist" learning environment. The course syllabus advises students to "think of them as Legos on steroids."

Chan took the class first, and Pondicherry the following year. The professor introduced them because of their similar approaches to the final project.

> "THE ORDERS MADE IT FEEL MORE REAL. LIKE, THIS IS ACTUALLY GOING TO HAPPEN."

"We both wanted to take it to the next level," Pondicherry recalls.

Chan's original name for the system — "LightUp"— is the one that stuck. Like LightUp, Pondicherry's "LogicBites" project had used magnetic circuit construction elements from the very first prototypes.

"There's just something magical and satisfying about magnets," says Chan.

"The magnets take away the 'almost' factor," Pondicherry adds. "It either connects

Malcolm Tyrrell

1. Denshi blocks are a circuit-building toy sold by Gakken.

2. Early LightUp prototype at Hack the Future 6.

3. In May, LightUp was chosen by a panel of venture capitalists as winner of the Pitch Your Prototype contest at MAKE's biannual Hardware Innovation Workshop.

Gregory Hayes

and works or doesn't. I think that encourages people to try things out instead of spending too much time thinking."

The pair had varying degrees of hands-on and research experience with other circuit-building toys.

"I actually didn't know about Braun Lectron (See "People Over Megahertz," page 12) or other systems like Denshi blocks," Chan explains, "until I started wondering what came before. I was more familiar with the RadioShack build-a-radio and 150-in-1 kits I played with as a kid. Now it's really interesting to look at how other systems have addressed the design constraints and compromises."

"I had researched many systems, both hardware and software, in an educational context," Pondicherry adds. "What we thought we could bring to the table was free-form play with basic components. And that's where the app came in."

LEARNING THE ROPES

To teach themselves manufacturing, LightUp applied to HAXLR8R, a hardware technology accelerator that works with startups in an intensive program based in San Francisco and Shenzhen, China. Among other things, the program allowed them to share office space with other HAXLR8R startups in Shenzhen.

There they built relationships with Chinese contract manufacturers and worked on LightUp's physical design, which has evolved from laser-cut, foil-taped prototype blocks to injection-molded plastic elements with stamped metal conductors.

"We're in our third major design revision," Chan says, "and working on the fourth now."

"If we include minor revisions," Pondicherry adds, "I would say 15-20."

AROUND THE CORNER

As with many crowdfunding success stories, LightUp had built up a large following before their Kickstarter even launched. That was back in May, when they won the Pitch Your Prototype contest at MAKE's Hardware Innovation Workshop. When the Kickstarter campaign closed, they had raised more than $120,000 — 240% of their $50,000 goal.

"All those orders made it feel more real," Pondicherry explains. "Like, this is actually going to happen."

Funding in place, Chan and Pondicherry are now intently focused on delivering on their promises. Orders for Kickstarter backers are slated to start shipping in December.

Eventually, they want to add logic gates, flip-flops, sensors, and microcontroller blocks to the system, as well as develop the app that guides users through building circuits.

"But," says Pondicherry, "running a business involves much more than just the technology."

Chan agrees: "We pretty much spend all our time just running around and doing what needs to be done." ◪

Andrew Terranova is an electrical engineer, writer, and robot hobbyist.

Illustration by Book Williams Jr.
See his unique process at
makezine.com/go/bookwilliamsjr

PICK YOUR BRAIN

A Vast Selection of Boards
Awaits Your Bidding

Small but powerful processors are the brains of maker projects, giving them electronic life. These mobile microcontrollers can be made to do virtually anything, from pinpoint motion control to advanced data tracking and analysis.

Starting with Arduino's recent mass adoption, programmable projects have become fast, cheap, and easy to produce. Makers have even begun designing their own new boards with sophisticated features like operating systems, GPS, and wireless communication; there now seem to be as many people creating these controllers as making projects using them. Our Field Guide will help you navigate the amazing assortment of microcontrollers and Linux-powered single-board computers for use in your next creation. Boot up — it's time to play.

WHICH BOARD IS RIGHT FOR ME?

Written by Alasdair Allan

A Field Guide To Microcontrollers and Single-Board Computers.

For a few months after Raspberry Pi came out, the choice was pretty simple. If you wanted to talk to arbitrary electronics, your best bet was to buy an Arduino microcontroller board; if you needed the power of an ARM-based processor to run Linux, the Raspberry Pi single-board computer (SBC) was the obvious choice (that is, if you could get your hands on one. Delivery issues are mostly resolved, but last year some people waited more than six months for their Pi).

Before Arduino and Raspberry Pi, things were more complicated. Going forward, things aren't just complicated again — they're bewildering. We're now seeing an explosion of new boards coming to market, and there's no reason to expect the trend to slow in the next year or two. If anything, I'm expecting more new boards to appear, not less — although most of them will disappear just as rapidly.

If you're old enough to remember the wide range of personal computers that sprung up in the early years of that industry — each with a different manufacturer, each based around a different CPU — then the state of the modern microcontroller board market may seem familiar. One looming question is whether, on the heels of this explosion in diversity, we'll see the rise of a monoculture, as we did in the desktop market. Or whether, perhaps, a more interesting ecosystem will emerge.

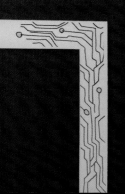

The
Intel
8008

Konstantin Lanzet

Life Before Arduino

The commercial microcontroller story starts, arguably, in 1971, with the arrival of the 4-bit Intel 4004. It was the second complete single-chip CPU in history, and the first to be commercially available. Its successor, the 8-bit 8008, would form the basis of the first personal computers.

Other processors from the era — like the Z80 that powered the TRS-80 in the U.S. and the Sinclair ZX Spectrum in the U.K., as well as the 6502 that powered the Apple II — are still around. Or at least their immediate descendants are still around, and now used in so-called "embedded systems."

But it was Microchip Technology's PIC microcontroller, dating from 1975, that became the backbone of the hobbyist market for many years, due to a combination of factors including low cost, ready availability, and a proliferation of free programming tools. The PIC is an MCU — a complete microcontroller unit — with on-board processor, memory, and programmable I/O.

Still available off-the-shelf today at less than $2 a chip, the PIC is a workhorse. Bare PIC microcontrollers can be a bit tricky to deal with if you're not used to low-level C programming, so Picaxe chips — standard PICs preprogrammed with firmware allowing them to "understand" BASIC or easy graphical flowchart languages — have become a popular way to use the PIC, especially in education.

While you can buy bare Picaxe chips, if you're new to the scene it's probably better to get a Picaxe starter kit designed to allow easy experimentation. In general these kit boards are made for prototyping and teaching rather than as bases for standalone projects.

The Parallax BASIC Stamp board — which is programmed in another variant of BASIC — is also a good alternative if you want to use the low-cost PIC microcontroller. Unlike Picaxe development boards, the BASIC Stamp is sold in standalone, single-board modules, like the more modern Arduino, that are intended to be the heart of a project. The BASIC Stamp also exploits the concept of add-on "carrier boards," like the Arduino "shield" system, except the Stamp sits on top of the add-on, rather than underneath. These carrier boards look a lot like Picaxe starter kits.

Development boards come in many different types and sizes, with vastly different feature sets. Comparing the particular spec of one board directly to another will rarely give you the full picture. Here are some of the common parts and terms that come into play when evaluating different boards.

1. Processor

The chip at the core of your board. It's the brain of your project, handling most of the functionality. Most processors can be classified as either a **microcontroller**, designed to control basic digital electronics, or a **system on chip** (**SoC**), which are more powerful processors similar to those in computers. Boards with SoCs are often called **single-board computers** (**SBC**).

2. Input/Output (I/O) Pins

Sockets that you use to connect LEDs, buttons, sensors, relays, motors, and other parts. You'll typically use a separate breadboard (a small panel with rows of sockets) and jumper wires to connect your board's I/O pins to electronic components while you're prototyping or developing your project.

The pins come in a few varieties, and often have multiple capabilities. **Digital** pins can read and control digital components. **Analog input** pins can read a range of voltages from analog components such as temperature sensors and dials. **PWM** pins allow for digital emulation of analog output. Some pins can also use communication protocols such as **serial**, **SPI**, **I2C**, or **CAN bus** to talk to other devices.

3. Power Input

For powering the board. Some boards accept a range of voltages, many others accept only 5 volts. Usually in the form of a DC barrel jack (pictured) or USB connector.

4. User LEDs and Buttons

Usable to indicate a status (in the case of an LED) or as an input (in the case of a button) without the need to wire up any additional circuitry. Other on-board LEDs may indicate whether the board is powered on, transmitting or receiving data, or accessing the flash memory.

5. Networking

On-board Ethernet ports, standard on most SoC boards and certain microcontrollers, allow you to connect to the internet via your router. Some boards even have built-in wi-fi chips for wireless connectivity.

6. USB Host Port

For connecting peripherals such as keyboards, mice, cameras, and wi-fi adapters. Available on many boards, especially those with an SoC processor.

7. Programming Port

Some boards connect to your computer via USB so you can reprogram the chip.

James Burke

Expansion Boards

Printed circuit boards that attach onto your development board to give it additional functionality such as Bluetooth, cellular, GPS, sound, graphics, and motor control. Also called **shields** with Arduino and **capes** with BeagleBone.

As microcontrollers and SoCs continue to advance, many expansion board functions are becoming incorporated directly into them.

Integrated Development Environment (IDE)

Where you write, compile, and debug your code. Many platforms use IDE software on your computer that also takes care of programming the chip, usually via its USB connector. Some network-enabled boards have a web-based IDE, so that you can use your internet browser to connect to the board and program it that way.

The boards' programming languages depend mostly on the platform. C, C++, Python, BASIC, and JavaScript are commonly used. SoC platforms are especially flexible and can be programmed in many different languages.

Libraries

Downloadable prewritten code that help you by making complex coding tasks much simpler. Frequently written to work with a particular board.

The March of the Arduino

Every so often a piece of technology can become a lever that moves the world, just a little bit. The Arduino is one of those levers.

It started off as a project to give artists access to embedded microprocessors for interaction design projects, but I think it's going to end up in a museum, someday, as a building block of the future world. Arduino allows rapid, cheap, prototyping for embedded systems. It turns what used to be fairly tough hardware problems into much simpler software problems. And it's become the poster child of the Maker Movement.

Based around the 8-bit Atmel AVR microcontroller line, the Arduino board breaks out digital, analog, and other pins from the controller in an idiosyncratic footprint that's become a de facto industry standard. It's a solid development platform, both for experienced hardware hackers and absolute beginners.

The real power of Arduino isn't really in the hardware, but the software — the Arduino IDE. While there are many other boards offering similar functionality, the Arduino has best succeeded at packing the complex, messy details of microcontroller programming in a user-friendly package. It has spawned many imitators and derivatives, and a huge community.

For now, at least, Arduino sits apart from the rest of the microcontroller market, and 20 or 30 years in the future we may look back on it like the Commodore 64, the Apple II, or (for the true old-timers) the PDP-11. These days Arduino is almost always a newbie's "first board," and it's influencing an entire generation of makers.

The Tessel

Gunther Kirsch

Microcontroller boards in the maker space have evolved to be easier to use and more accessible, and a lot of that can be directly attributed to Arduino and its imitators. Their model for software development has been copied many times, and the Tessel stands out as an interesting departure.

Though it takes a different approach, the Tessel is really a continuation of the Arduino idea — boards that can be programmed in ways that are familiar to software rather than hardware developers. The Tessel's operating system is a JavaScript interpreter built around the Lua runtime, and is compatible with the node.js API — effectively an event loop on bare metal. It promises to take advantage of the sprawling node.js community and will come with built-in wi-fi.

While slow compared to a contemporary JavaScript engine, the overhead of the Lua runtime they're using is small — kilobytes not megabytes — which means it can run on a $3 ARM Cortex-M3. The Tessel isn't meant as a competitor for the Raspberry Pi and other relatively "heavyweight" Linux-based SBCs. It's about scaling down, not up.

The Tessel will ship with an Arduino extension board for plugging in shields. It should even be able to use Arduino libraries and run sketches directly. It's a board designed from the ground up to be part of the Internet of Things.

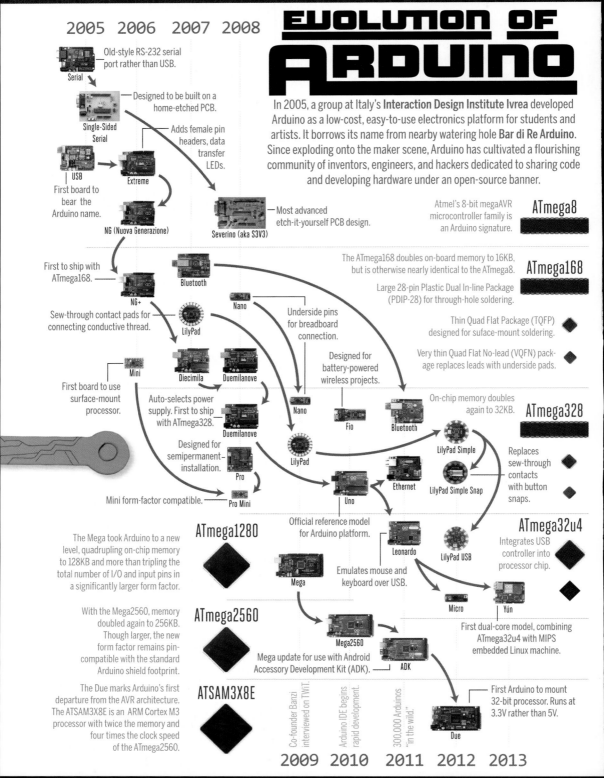

EVOLUTION OF ARDUINO

2005 2006 2007 2008

In 2005, a group at Italy's **Interaction Design Institute Ivrea** developed Arduino as a low-cost, easy-to-use electronics platform for students and artists. It borrows its name from nearby watering hole **Bar di Re Arduino**. Since exploding onto the maker scene, Arduino has cultivated a flourishing community of inventors, engineers, and hackers dedicated to sharing code and developing hardware under an open-source banner.

Serial — Old-style RS-232 serial port rather than USB.

Single-Sided Serial — Designed to be built on a home-etched PCB.

USB — First board to bear the Arduino name.

Extreme — Adds female pin headers, data transfer LEDs.

NG (Nuova Generazione)

Severino (aka S3V3) — Most advanced etch-it-yourself PCB design.

Atmel's 8-bit megaAVR microcontroller family is an Arduino signature. **ATmega8**

First to ship with ATmega168. — **NG+**

Bluetooth

Nano

The ATmega168 doubles on-board memory to 16KB, but is otherwise nearly identical to the ATmega8. **ATmega168**

Large 28-pin Plastic Dual In-line Package (PDIP-28) for through-hole soldering.

Sew-through contact pads for connecting conductive thread. — **LilyPad**

Underside pins for breadboard connection.

Thin Quad Flat Package (TQFP) designed for suface-mount soldering.

Mini — First board to use surface-mount processor.

Diecimila   **Duemilanove**

Designed for battery-powered wireless projects.

Very thin Quad Flat No-lead (VQFN) package replaces leads with underside pads.

Auto-selects power supply. First to ship with ATmega328. **Duemilanove**

Nano

Fio

Bluetooth

On-chip memory doubles again to 32KB. **ATmega328**

LilyPad Simple

Designed for semipermanent installation. **Pro**

LilyPad

Replaces sew-through contacts with button snaps.

Mini form-factor compatible. — **Pro Mini**

Ethernet

LilyPad Simple Snap

Uno

Official reference model for Arduino platform.

ATmega1280

The Mega took Arduino to a new level, quadrupling on-chip memory to 128KB and more than tripling the total number of I/O and input pins in a significantly larger form factor.

Mega

Leonardo

Emulates mouse and keyboard over USB.

LilyPad USB

ATmega32u4

Integrates USB controller into processor chip.

With the Mega2560, memory doubled again to 256KB. Though larger, the new form factor remains pin-compatible with the standard Arduino shield footprint.

ATmega2560

Mega2560

Micro

Yún

First dual-core model, combining ATmega32u4 with MIPS embedded Linux machine.

Mega update for use with Android Accessory Development Kit (ADK). — **ADK**

The Due marks Arduino's first departure from the AVR architecture. The ATSAM3X8E is an ARM Cortex M3 processor with twice the memory and four times the clock speed of the ATmega2560.

ATSAM3X8E

First Arduino to mount 32-bit processor. Runs at 3.3V rather than 5V.

Due

Co-founder Banzi interviewed on TWiT.

Arduino IDE begins rapid development.

300,000 Arduinos "in the wild."

2009 2010 2011 2012 2013

The latest board in the series, the Arduino Leonardo, differs from its predecessors in that, in addition to the virtual serial port necessary to transfer code from the IDE to the board, it can also appear to a connected computer as a USB mouse and keyboard.

Alternatives to Arduino

The Arduino-and-derivatives phenomenon has driven interesting innovation, and convergence, in the microcontroller marketplace.

The LaunchPad MSP430

The Texas Instruments MSP430 is very similar to the Atmel ATmega microcontroller chip. Notable differences include a very low price point, as well as some interesting refinements for low power consumption. It's also readily available in the through-hole DIP form factor, while dual-inline-packaged ATmega chips often seem to be in short supply. If through-hole mounting is important to you, take a look at the MSP430. The easiest way to get acquainted is to pick up a TI LaunchPad developer board.

The major difference between LaunchPad and Arduino is cost. While a new Uno will run you $30, and a Leonardo $25, the LaunchPad MSP430 rings up at just $10 directly from

The Arduino Uno

TI LaunchPad

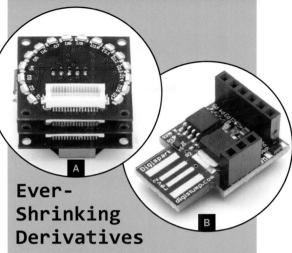

A

B

Ever-Shrinking Derivatives

As discussed, the success of the Arduino has led to numerous copies and compatible boards arriving on the market. The crowdfunding site Kickstarter is littered with them, some amazingly successful, some not so much. It'd be impossible to list them all, but there are some that stand out, chiefly because of their size (or lack thereof).

The TinyDuino, for example, is an Arduino-compatible microcontroller using the same processor as the Arduino Uno, but at the size of a U.S. quarter (**Figure A**). The main processor board includes the microcontroller and supporting circuitry, while the USB and DC power regulators (among other things) have been offloaded to shields. If you don't need them for your project, you don't have to install them. However, despite its size, or more probably because of it, the TinyDuino costs $20 for the main processor board, plus another $18 for the USB/ICP programmer shield you're likely to need. Miniaturization doesn't come cheaply.

The DigiSpark is another tiny Arduino-compatible board (**Figure B**). It is built around the ATtiny85 microcontroller, making it much less powerful than the TinyDuino. It only has 6 I/O pins but, on the other hand, it costs just $9. Like the TinyDuino, it has a variety of interesting shield kits allowing you to easily extend its capabilities.

The Picaxe-28X2 Shield Base

TI or a major distributor (and that includes a USB cable). The upcoming USB LaunchPad MSP430 adds on-chip USB for only $2 more. I've seen LaunchPad boards available for less than $5.

Although the MSP430G2553 chip, the version that's used by the LaunchPad, has only 14 I/O pins and 16K of program memory in comparison to the Uno's ATmega328 (with 32K and 20 I/O pins), this may well be all you need for your particular project.

Until recently, the MSP430's programming environment was a bit of a hitch. For a generation of makers accustomed to the user-friendly Arduino IDE, the MSP430's old Eclipse-based development environment seemed overly complicated and hard to use. The new, open source Energia prototyping platform has changed all that. With cross-platform support for Windows, OS X, and Linux, it brings the Wiring and Arduino frameworks to the MSP430 in style. Energia lets you take your Arduino source code — your sketch — and simply drop it directly onto the MSP430.

Of course a lot of the power of the Arduino is in its community code libraries, but a lot of these have been ported over. Unless you need something fairly obscure, the arrival of Energia means that you can use the TI LaunchPad almost exactly as though it were an Arduino.

Picaxe Strikes Back

The near-ubiquity of the Arduino platform has driven many systems that aren't software-compatible to be at least physically compatible.

The Picaxe-28X2 Shield Base replicates the

Arduino form factor, establishing compatibility with the hundreds of Arduino shields already on the market.

Wiring

The runaway success of Arduino has stolen some love and attention from the Wiring board and its programming environment. It probably deserves more.

The board is based around the same Processing-derived development environment from which Arduino came — although by now it's a somewhat different branch of the family tree — and should feel familiar to anyone used to the Arduino IDE. It's possibly just different enough, however, to trip you up if you get careless.

The Wiring programming environment supports any hardware based on the Atmel AVR series of processors, not just the Wiring boards themselves. And this includes the Arduino line.

As of this writing, support for the AVR XMEGA, the tinyAVR, the TI MSP430, the Microchip PIC24/32 Series, and the STM M3 ARM cores is billed as "coming soon." Delivery on that promise, if it happens, will be a very interesting development, as it will allow Arduino-compatible code to be deployed onto a range of microcontroller architectures.

The latest board, the Wiring S, is similar to the older Arduino Diecimila, but with a bigger processor. Like the Picaxe Shield Base, pairing the board with a Wiring S Play Shield makes it pin compatible with the Arduino form

The Wiring S board

The Netduino Plus

factor so you can reuse your Arduino shields with your Wiring boards.

The Netduino

The Netduino, too, adopts the Arduino form factor — you can attach most existing Arduino shields. But that's where the similarity ends.

There are several Netduino boards available, and unlike the boards we've met so far that have all been based around 8- or 16-bit microcontrollers of one type or another, the netduino is an ARM Cortex-based board built around the 32-bit STMicro STM32Fx microcontroller.

The operating system on the board is the .NET Micro Framework. These boards are programmable in C#, directly from Microsoft Visual C# Express 2010, and are extremely powerful and flexible. C# developers on non-Windows platforms aren't entirely out in the cold, as there is some support for OS X and Linux.

The Parallax Propeller

This is an interesting alternative to other microcontroller chips on the market. Where almost every other one has a single processor core, the Propeller has eight.

That means eight separate processes can be running simultaneously, monitoring and responding to sensor and other inputs. Think about it as eight simultaneously-running Arduino `loop( )` functions.

Depending on your application, running parallel processors instead of using interrupts can be surprisingly powerful, and at $50, it's not much more expensive than the other

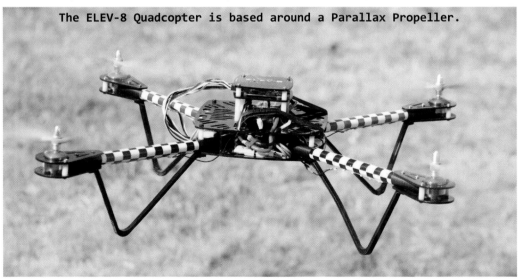

The ELEV-8 Quadcopter is based around a Parallax Propeller.

The Arduino Yún

The BLEduino

microcontroller boards we've talked about.

While the Propeller is available in a variety of form factors — including a bare chip in both DIP and SMT (Surface-Mount Technology) packages, if you want that for prototyping — like many other manufacturers Parallax has adopted the Arduino form factor for its Parallax Propeller ASC+ board.

Going Wireless

There is a sea change going on in the microcontroller world: Everything is going wireless. A range of shields is available for Arduino and Arduino-compatible boards boasting GSM cellular, wi-fi, Bluetooth Low Energy, and other wireless capabilities.

Wi-Fi

Announced in May at Maker Faire Bay Area by Massimo Banzi himself, the Arduino Yún is the first in a series of embedded Linux boards to bear the Arduino name, and it comes with integrated wi-fi.

The board is fundamentally an Arduino Leonardo, sporting an ATmega32U4 microcontroller, plus a separate embedded AR9331 processor running a MIPS Linux variant based around the OpenWRT distribution. You can program it remotely via wi-fi or with the usual USB cable. Suggestively, perhaps, they've also partnered with Temboo for one-stop API access to data from Twitter, Facebook, Foursquare, FedEx, PayPal, and more.

The board should come in at $69, which isn't bad when you consider the price of bundling an embedded Linux board, an Arduino, and a wi-fi dongle or shield all together.

Bluetooth Low Energy

The arrival of BLE has changed the playing field for wireless in embedded devices. Fixing most of the (numerous) problems with the old standard, the new Bluetooth LE protocol is much easier to work with than "classic" Bluetooth. While some smartphones, including the iPhone, have had Bluetooth LE support on board for awhile, there has been a delay in getting support for it out into the hands of makers. About six months ago, boards like RedBearLab's BLE Shield and BLE Mini started to arrive, and now we're seeing a spate of Arduino-compatible boards with integrated Bluetooth LE.

Among several recently active Kickstarter projects working to produce such a board, two of the most widely discussed are BLEduino and RFduino.

Interestingly, both are small form-factor boards, which is indicative of their intended uses.

Mesh Networking

If you need to cover a large geographical area with a wireless network, mesh networking is an ideal solution. Each board talks to every neighboring board, transferring packets across the ad-hoc network to the edge, where there is a router or a gateway out into the wider world and the internet.

The Geogram One

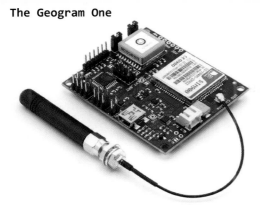

Another runaway success on Kickstarter, the Pinoccio is an Arduino-compatible board with built-in 802.15.4-based mesh networking and a LiPo battery, with additional wi-fi capability via a shield (see *"The Tale of Pinoccio,"* page 66). It looks like a perfect platform to build a distributed sensor network with very little effort.

GSM

The Geogram One is an Arduino-compatible board intended for tracking applications. It has both a GSM cellular modem and an on-board GPS receiver. Despite that, it's still an Arduino under the hood, with all the flexibility that implies.

Want to see more?
Find our favorite Raspberry Pi-controlled projects at
makezine.com/go/rpi

The Arrival of the Raspberry Pi

The single-board Linux computer, existed well before the arrival of the Raspberry Pi; I personally was using Gumstix boards fairly extensively about 10 years ago. Recently, however, like the Arduino before it, the Raspberry Pi has single-handedly rebooted the market, this time for single-board computers. Also like the Arduino, it has brought an explosion of would-be competitors.

Unlike the Arduino, the Raspberry Pi was never really designed as a platform for makers. But the rock-bottom $35 price created a large market for single-board computers almost overnight, and it was months after the official release before the Raspberry Pi supply caught up with demand.

It was designed from the start as a low-cost platform for kids to learn programming — a cheap educational tool. Despite that, rather than because of it, thousands of creative computer-embedded projects are being built around the board. Like with the Arduino, it's the Pi's flourishing community that has made it a success.

The Raspberry Pi

Gunther Kirsch

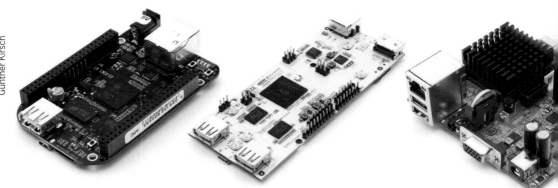

The BeagleBone(s)

While expensive at $89, TI's BeagleBone was designed from the ground up to talk to arbitrary bits of hardware — sensors, actuators and other electronics. It was a board designed from the start for makers, rather than as an education platform like the Pi.

Unfortunately the price break with Pi was just too tempting for most buyers; while the BeagleBone developed a small but dedicated following — for instance the Ninja Block system was built around BeagleBone — it was never really a rival for the Raspberry Pi.

That all changed with the arrival of Beagle-Bone Black. Besides the new color, the newer board looks pretty familiar; it has the same footprint as the original BeagleBone, and a similar layout. Among other interesting new features, BeagleBone Black moves the operating system off the SD card to on-board flash memory, freeing up the microSD card slot for other purposes.

Most crucially, however, the board has dropped in price from $89 to $45, which is real competition for the Pi's $35 retail price point. Especially when you consider the Black's better specs, greater flexibility, and generally better user experience.

The pcDuino

The pcDuino is another embedded board running Linux. Interestingly, though it is Arduino pin-compatible, it is not form-factor compatible. All the pins you'll need to use most Arduino shields are "broken out" of the ARM

Cortex-A8 processor.

You can write code directly on this board, as if it were an Arduino, then run it natively on the board. SparkFun is even in the process of putting together an adaptor to make the board footprint- and pin-compatible with Arduino. At $60, it's an attractive option, and looks easy enough to set up.

The x86 Fights Back?

Single-board computers running Linux have traditionally made use of ARM processors; it's only recently that x86 boards have started to appear. Perhaps the best example of this is AMD's Gizmo Board.

Essentially a laptop on a single board, Gizmo is blindingly fast, extremely flexible, and hugely powerful by the standards of those of us coming from the microcontroller world. But that performance does come at a price — at $200, it's not cheap.

Hybrid Boards

Today there's a proliferation of boards that seem to want to be all things for all users, combining a Raspberry Pi-like SBC with an Arduino-like microcontroller. In addition to the Arduino Yún (see "Wi-Fi, page 56), the Udoo, for instance, made a huge showing on Kickstarter. It's an ARM-based Linux board like the Raspberry Pi, built around an impressive dual- or quad-core ARM Cortex-A9 CPU, with a second ARM processor, the SAM3X, alongside to mimic the Arduino Due. It's priced to match its performance at $130.

The Udoo

Around the Corner

For a couple of years, there was a phenomenon among programmers where everyone's first "serious" application was a Twitter client. Before Twitter existed, everyone seemed to write a text editor the first time out. Why? Everyone used a text editor, and later a Twitter client, and so everyone had an opinion about how they should work. Among existing programs, the buttons weren't in just the right places or the workflow was slightly wrong. So everyone just wrote their own — they scratched their own itch.

I think that's exactly what's happening with the current explosion of Arduino-compatible boards on Kickstarter. Everyone uses the Arduino, but everyone uses it for slightly different purposes. So as their first "serious" hardware project they decide to build their own version and scratch that itch.

I fully expect that many of these boards will disappear after one short run, for the same reasons that most of those new Twitter clients quickly disappeared: The costs of supporting them will greatly exceed the income they generate.

But then, a lot of very serious people in the open source world got their start writing text editors or Twitter clients, and like them, many of these fledgling pro-makers might well go on to do much more interesting things than designing Arduino derivatives.

In circuit boards, the gap between concept and production-ready prototype is shrinking so quickly right now that it's very hard to

Wearables

Starting in 2007 or so, the phrase "wearable microcontroller" was pretty much synonymous with the LilyPad Arduino, a system of sewable electronic modules designed by Leah Buechley.

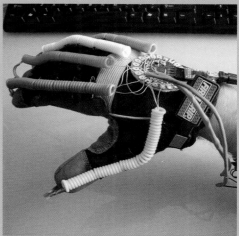

Jacek Spiewla

Jacek Spiewla's BeatGlove is a wearable electronic musical instrument based on LilyPad Arduino.

In 2012, Adafruit Industries introduced the first major LilyPad competitor — the Flora. Though the Flora is designed to be more beginner-friendly than the LilyPad, the two platforms are still fairly closely matched. Adafruit promises a second smaller, wearable board later this year, and it looks like the wearables space, which has been static for a few years, is going to start heating up.

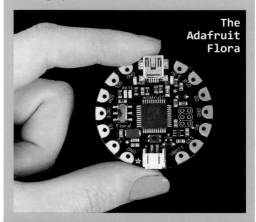

The Adafruit Flora

Becky Stern

anticipate what the next big thing will be. But the prevalence and proliferation of wireless microcontrollers is, I think, giving a big hint.

Everyday objects are already becoming smarter; in 10 years, every piece of clothing you wear, and every object you carry on your person, will be measuring, weighing, and calculating. By then, the world will be full of sensors, and those sensors will need to talk to each other.

Because of the communities that have grown up around them, I would unhesitatingly recommend an Arduino if you need an 8-bit microcontroller, or a Raspberry Pi if you need a single-board computer running Linux.

If you're leaning toward the Pi, but worried it may not suit your application, the decision gets more complex. The Raspberry Pi is yet to become an unstoppable force, or an immovable object, like the Arduino. The most serious alternative, around the same price point, is BeagleBone Black. On the other hand, BeagleBone Black is relatively new, and its community is much smaller, so you might end up having to solve a lot of your own problems.

If you're leaning toward an Arduino, but have specific needs (like wireless connectivity) that it doesn't meet out of the box, then you should probably look first among the myriad of Arduino derivatives. You'll probably find your desired feature set baked right into one of them.

Finally, if your project's I/O requirements permit it, take a serious look at the TI LaunchPad MSP430. Its low-price, low-power requirements, and user-friendly development environment make a very strong case. ∎

FPGAs

FPGAs (field-programmable gate arrays) represent an entirely new class of boards. With microcontrollers, you have control over the software, the code living on the chip. With an FPGA, you start with a blank slate, and *design the chip itself* at the hardware level. There is no processor to run software until you make it happen.

It might sound crazy, but it gives you flexibility. If you need more than one serial port then you just add another to your chip design. It also means that you can design the hardware to be a processor that you can write software for. Many companies, like Intel, use FPGAs to prototype their chips.

The Gadget Factory's Papilio One is an open source FPGA project board intended for hobbyists and newcomers. Based on the Spartan 3 FPGA chip, it's got 48 I/O pins and comes with two Arduino-compatible "soft processors" ready to load into the array, so you can quickly get up and running using the Arduino IDE. Starting at $38, it's a solid entry-level FPGA. For something more capable, look at the Papilio Pro ($85) or Embedded Micro's Mojo board ($75), which gives you a Spartan 6 chip, 84 I/Os, 8 analog inputs, and 9 on-board LEDs.

The Papilio One FPGA Development Board

Slightly more expensive than the Papilio One, Embedded Micro's Mojo board is a correspondingly more capable platform.

Alasdair Allan is a scientist, author, hacker, tinkerer, and co-founder of the Thing System, a startup attempting to fix the Internet of Things. He spends much of his time probing current trends to determine which technologies are going to define our future.

NOW WHAT, NOOBS?

Written by Craig Couden

Four fun and easy Arduino projects to spark your microcontroller skills.

Now that you've gotten a primer on the world of microcontrollers, you probably want to try assembling an electronics project of your own. Here are a few easy Arduino creations to get you started — find these and more at makezine.com/projects.

Monkey Couch Guardian

makezine.com/projects/monkey-couch-guardian
This obnoxious device discourages cats and other fur-shedding pets from jumping on beds and couches. It uses a simple PIR (passive infrared) sensor circuit to control the battery-powered, cymbal-banging monkey. It'll also adequately annoy your parents/roommates/kids.

Wii Nunchuk Mouse

makezine.com/projects/make-33/wii-nunchuk-mouse
Add motion control to your PC by converting a Wii Nunchuk into a mouse. Its native I²C serial protocol is easy to interface with Arduino, and as a bonus the connector will accept standard jumper wires, so there's no need to cut up the cable or use a dedicated adapter.

Fade Colors on an RGB LED

makezine.com/projects/use-a-common-anode-rgb-led
Ready to add some sweet mood lighting to your next project? This quick tutorial shows how to fade between colors with an Arduino and common RGB LED. Use it to add pizzazz to your holiday adornments, or put it inside your Millennium Falcon model.

Control a Servomotor

makezine.com/projects/control-a-servo-with-a-force-sensitive-resistor
This simple introduction uses force-sensitive resistors to control servomotors. The servo's position will depend on the force reading from the sensor — great for DIY animatronics or pressure-sensitive practical jokes. ◾

Build one of these, or another project you want to share? Let us know!

THE BOARD ROOM

Meet nine new boards and the cool projects you can make with them.

The number of exciting new microcontrollers hitting the market is growing daily, incorporating features like wireless communication and built-in motors. We asked the teams behind nine of the latest boards to each share a favorite project that uses their device.

1. Mojo Easy-to-use FPGA development board.
Giant Graphic Equalizer

To demonstrate the power of the Mojo board (embeddedmicro.com), we built a huge (2.5'×1.25') equalizer consisting of three sheets of laser-cut acrylic forming a 10×10 grid of rectangles, each with three RGB LEDs for a total of 900 LEDs (pictured above). The LEDs are driven with MOSFETs that are directly controlled by 70 of the 84 digital IO pins the Mojo features. The equalizer also has an onboard microphone that is connected to one of the Mojo's analog inputs, something not often found on FPGA boards. To perform live audio visualization, the Mojo continuously samples the microphone, storing samples in a buffer. Once the buffer is full, the samples are fed into an FFT that performs the frequency analysis. The output from the FFT is then used to generate a frame for the display in full 24-bit color and a frame rate of roughly 190 frames per second. The display is double buffered and synced to prevent any artifacts. All of this is accomplished using only roughly 20% of available space in the Mojo!

2. TinyDuino Arduino-compatible, smaller than a quarter.
GPS Cat Collar

Our 9-year-old male cat Conley loves to roam around the neighborhood for hours at a time, and we have no idea what he's doing or where he's gone — until now. We decided to create a GPS tracking device for him using a TinyDuino (tiny-circuits.com), a miniature Arduino compatible board that's smaller than a quarter and extremely light. This GPS device records the position and logs the data every second to an attached microSD card. When he gets back to the house, we can pull out the microSD card, put it in our computer, and check out where he went using Google Maps.

3. Spark Core   Wi-fi for internet-connected hardware.

Simple Security System

Why pay $49 a month for a security system when you can do it yourself? We paired a Spark Core (sparkdevices.com) with a PIR (passive infrared) motion sensor to create a simple security system that generates an internet "event" whenever motion is sensed. By pairing this project with Twilio, the cloud communications service (think APIs for SMS text messages), we made a system that sends you a text message every time it detects motion. Take it one step further to add geolocation information from your phone, and you could make it text you anytime motion is sensed while you're away from home.

4. Moti   Smart motor with Arduino-compatible board.

Smartphone-Controlled Turntable

We wanted to build a turntable that can be scratched from a tablet or phone. Our programmable smart motor, Moti (moti.ph), makes this pretty straightforward, as it has an Arduino-compatible micro, motor driver, and continuous encoder built in. We laser-cut the body out of plywood and added some audio electronics pilfered from an old record player. A Bluetooth shield was attached directly to the breakout pins on the Moti, and it was slotted into place. We fired up the Moti app, which automatically detects the motor and offers graphics to control its speed and position. Just spin the dial, and the motor follows. Our friend Rob designed a custom interface using JavaScript to control the turntable through Moti's RESTful API, and we had a turntable you can scratch with a virtual record. DJ Moti in the house!

5

6

5. UDOO For Android, Linux, Arduino, and ADK 2012.

High-Quality Music Player

Tsunamp is a free and open source Linux distribution that transforms UDOO (udoo.org) into a high-quality music player. It's designed to achieve excellent sonic results in a user-friendly environment. Basically, when you flash Tsunamp into UDOO, it becomes a standalone audio player which can then be controlled using mobile phones, PCs, Macs, and tablets, thanks to UDOO's built-in wireless adapter. It can retrieve a music library stored on a NAS or a USB drive, play web radio, and even act as an airport receiver. Grab it for free at tsunamp.com.

6. Digispark Cheaper/smaller than Arduino, with ATtiny85.

Bluetooth-Controlled Robot

Many makers use the Digispark (digistump. com) for projects that communicate with a host computer, because it is able to emulate a keyboard, mouse, joystick, or send data directly. One of the more complex and memorable standalone projects built with the Digispark is the CamBot by Dave Astolfo. The CamBot is a Bluetooth-controlled robot built from Legos, a Digispark, a Digispark Motor

Driver Shield, a cheap Bluetooth module, a wi-fi webcam, and some motors. The bot is used to check hard-to-reach places, such as heating ducts, and it can be controlled from a smartphone, allowing the driver to control it and see the video feed. Dave did more with a single Digispark than most projects that use full-blown Arduinos. We can't wait to see what our users come up with next!

7. JeeNode Arduino-compatible, with Atmel 8-bit RISC.
Networked Energy Monitors

The Western Cooling Efficiency Center at University of California, Davis, used the low-power JeeNode (jeelabs.com) prototyping kit to build groups of networked devices (communicating via the built-in 915 MHz radios) to motivate and monitor simple energy-saving actions in students' apartments. The devices were able to run off two AA batteries for the two-month study without dying!

8. BLEduino Arduino-compatible with Bluetooth Low Energy.
Smartphone Game Controller

In the Virtual Controller project, the BLEduino (bleduino.cc) is used to play classic video games on the computer by serving as a receiver for a virtual controller on your phone. When the iPhone app registers a button tap it sends a command to the BLEduino, which then maps it to keyboard strokes on the computer. (The BLEduino app is a really important aspect of our board since it provides all of the UI and libraries necessary to communicate with the BLEduino.) The BLEduino creates the keystrokes by simulating a normal keyboard hooked up to the computer via the USB port and receives the commands via BLE.

9. TinyG Multi-axis motion control system.
Computer-Controlled Three-Axis Mill

TinyG (synthetos.com) is a complete embedded multi-axis motion control system on a single board. It makes industrial-grade control affordable and accessible while still being powerful enough for professionals. It's used in pick-and-place machines, small industrial production lines, and other applications that require precise motion control. Othermill (otherfab.com/products) is a portable, computer-controlled three-axis mill that uses the TinyG controller. Othermill is precise enough for detailed electrical and mechanical prototyping work, yet compact and quiet enough for home use.

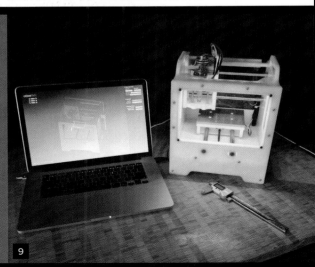

ON THE WEB: *Facilitating hardware creation with Seeed Studio's Eric Pan:* makezine.com/go/seeedstudio

THE TALE OF PINOCCIO

Two makers' journey of starting a new company.

Written and illustrated by Sally Carson

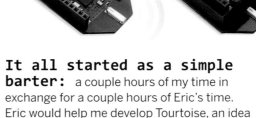

It all started as a simple barter: a couple hours of my time in exchange for a couple hours of Eric's time. Eric would help me develop Tourtoise, an idea I had for a bicycle-touring app. And I would help Eric with his sensor project.

Eric Jennings and I recently celebrated our company's first birthday. Today, we refer to Pinoccio (pinocc.io) as "a complete hardware + software ecosystem for building the Internet of Things." Looking back though, we didn't start this journey with such a clear vision of the product. This is the story of Pinoccio's first year — the blind alleys, missteps, insights, and epiphanies. Let's go back in time.

During our very first conversation, Eric told me about what a hard time he was having getting his DIY projects to talk to the web wirelessly. He's a brilliant dude, with a strong mix of both software and electrical

engineering. Someone with his expertise having this problem signaled an opportunity — a probletunity! To address it, Eric wanted to develop a wireless, remote sensor network. Of course, this would eventually become much more than just a sensor network.

I told Eric that the project reminded me of Bruce Sterling's concept of "spimes" from his 2005 book *Shaping Things*. Spimes, as Sterling describes, are objects that can be tracked in space and time. They are physical manifestations of what is primarily digital information, or as he puts it, "material instantiations of an immaterial system." The book essentially describes the Internet of Things, without ever referring to it as such.

Eric read the book, and I reread it. In fact, this was probably my seventh time reading it. Much of what Sterling described was theoretical in 2005, but had since materialized. It all felt very futuristic! I was so excited to be involved with this project, originally part of a simple barter, that I soon told Eric "You know, I like your idea better." And with that, Tourtoise climbed back into his shell (sorry, Tourtoise).

Part of what had me so fired up was Sterling's idea that spimes would be instrumental in preventing the most devastating effects of climate change. The idea being that we may not know what data we'll need until we need it, so let's start instrumenting everything now so we have a rich backlog when a need does arise.

This idea is reminiscent of the SafeCast project (safecast.org), where makers gave DIY Geiger counters to Japanese citizens after

the Fukushima nuclear disaster. Empowering citizens with access to accurate data about the world that they live in is one of our main goals with Pinoccio.

But, back to the technical problem that we were trying to solve: getting wireless hardware to easily talk to the web. The applications for Pinoccio seemed endless, so we felt we should try to focus on a pilot project. We selected automated sprinkler systems as a way to test the technology.

We spent the next month or two digging into the world of automated irrigation. Eric and I have both spent our careers working on the web, so we weren't exactly in our wheelhouse. But we tried to learn as much as we could. We met with experts. We researched soil moisture sensors. Eric even toured the Desert Research Institute in Nevada. We also learned about what motivated people to conserve water (turns out, giving them data that compares their usage with their neighbors works well).

We discovered that automated irrigation was a huge opportunity, and we started to

feel like we could build a successful business on this idea. On top of that, the work was interesting, it was challenging, and we felt that water conservation was a worthy cause. But still, something wasn't quite right. Sprinklers just weren't our passion.

We had lost the scent of what had originally gotten us fired up about Pinoccio. We had been distracted by what I call "shiny things." In this case, the shiny thing was the potential to create a viable business around sprinklers. Now we had to pause and recenter. What was the problem we were most interested in solving?

We thought back to what had gotten us so excited in the first place: We wanted to build stuff with Pinoccio! We wanted to make bike computers and quadcopters and DIY Fitbits. And when we described the system to our buddies, they had even better ideas

for how they'd like to use it; like getting a text message if their frog research lab got too warm or creating a city-scale mesh network for civic hacking. Practically every time we described Pinoccio to someone, a new brilliant idea for its use emerged.

By zooming out, we realized that we didn't want to build a product for consumption — we wanted to build a platform for creation. A DIY Internet of Things! This was the idea that had us jumping out of bed in the morning, rushing to get to work. We moved back on track with a renewed sense of enthusiasm and momentum.

Now that our vision for Pinoccio was starting to take shape, we needed to figure out if this was something that other people wanted too. We identified who we thought our earliest customers would be: makers and hardware-curious software developers. We attended MAKE's Hardware Innovation Workshop, the Open Hardware Summit, and World Maker Faire in order to observe and interview our target audience.

The goal was to learn about their "emotional requirements" — their needs, motivations, and goals. We were careful not to ask leading questions like "Would you buy Pinoccio?" Instead we asked open-ended questions about actual past behaviors like "What's the coolest thing you've ever built?" and "What tools do you use?" We took copious notes, which we reviewed back at the office.

The interviews confirmed that we were onto something that people needed. The insights helped shape early product decisions and requirements. Long battery life was critical, and wi-fi is power hungry, so the microcontrollers (we call them "Scouts") would speak to each other over low-power mesh radio (802.15.4). A single Scout would connect the entire group of Scouts (the "Troop") to the web via a wi-fi

Backpack (yep, we call shields "Backpacks"). Finally, a new Troop must talk to the web wirelessly within moments of unboxing.

About four months of prototyping and refining later, we decided we were ready to launch a crowdfunding campaign. We figured if we couldn't reach our funding goal, then we had missed the mark and it was back to the drawing board. To our delight, the campaign was a success. We got wonderful feedback, met amazing makers from around the world, and a community of eager campaign funders started to coalesce on our forums.

As I write this, we're in the throes of producing our first run of boards. We've learned so much in this first year. We learned to listen to our own hearts about working on what's most important to us. We learned to listen to our customers about what they really need. Now, we can't wait to send Pinoccio out into the world to see what amazing things you'll build with it! ☑

Sally Carson is user experience designer and co-founder of Pinoccio. She seriously won't shut up about bikes, and she has a comic book series about her experiences as a NYC bike messenger called *The Skids*.

5 Cool Things You Can Do with Pinoccio

1. BUILD A WEBROVER.
We hooked up Pinoccio to a Pololu 3pi robot to create an unmanned rover that can be driven via the web. Then we drove our little WebRover in Nevada from 10,000km away in Brazil. Check out the video and the project code: pinocc.io/examples/webrover

2. TXT ME IF _____ GETS TOO HOT/COLD.
Science labs, kegerators, small batch chocolate. Lots of things need to stay within a certain temperature range. Get notified if your precious project gets too hot or too cold.

3. OPEN, SAYS ME!
Hook a Pinoccio Scout up to your bike and have your garage door automagically open as you arrive. Your neighbors will wonder if you're a super spy.

4. MAKE 3D PRINTED OBJECTS SMART.
That blobject your Replicator just spit out sure is cool. But wouldn't it be cooler if it had a brain? Leave a hole for charging via micro USB, or attach the Solar Backpack and trickle charge off any visible light, even a desk lamp!

5. MAKE A REVOLUTION COMMUNICATION DEVICE.
Worried your cell network or Twitter access is going to get blocked during a peaceful protest? Spin up your own mesh network of individually addressable devices for communicating with each other during your next demonstration.

CLOUDFRIDGE

Put your icebox on the Internet of Things.

Written by Tod E. Kurt and Mike Kuniavsky

Gunther Kirsch (photo); Fonco Creative and MAKE (diorama)

The Internet of Things (IoT) is essentially the internet we already have, but now it's not just humans using it — all kinds of devices are using it too. Today you can buy a washing machine that tweets when it's done or an irrigation controller that checks the weather before watering.

Free data hosting services, and Arduino libraries for accessing them, make it easy and inexpensive to connect virtually any sensor to the cloud and read data back. These services store historical data, provide analytics tools, and reformat data streams so they're easy for other devices to read.

The infrastructure is here for experimenting with your own IoT devices. So, what do we want to connect to the internet, and why?

CloudFridge

Your fridge is an electricity hog, and every time someone opens the door, the cold air pours out, warms air slides in, and the fridge has to cool it again. Opening the door is responsible for 7%–10% of the cost of running a fridge, burning about 28 watt-hours (Wh) every time, or 500 kilowatt-hours a year — $100 or more annually. Leave it open too long, and the compressor runs continuously, using about 1kW per hour. Perhaps now those midnight snacks won't seem as appetizing.

This project connects your fridge door to the Xively data service using an Ethernet-connected Arduino and a magnetic sensor. It monitors how often the door is opened and for how long, and it creates a persistent data archive you can use to detect usage trends. Xively is free for small projects, and it was started by friends of ours. But there are many similar services such as Nimbits, Paraimpu, ThingSpeak, 2lemetry, sen.se, and ioBridge.

1. Set up your data feed.

Create an account at xively.com. Click on the Develop tab, then select Add Device. Enter the name and description for your CloudFridge device, choose the Private Device setting, and click Add Device.

MATERIALS:

» **Arduino Uno microcontroller board** Maker Shed item #MKSP11, makershed.com, $30
» **Arduino Ethernet Shield** Maker Shed #MKSP7, $60
» **BlinkM Smart LED** Maker Shed #MKTNC1, $15
» **Reed switch** from a mini door/window alarm, such as Amazon #B008NXFKLK
» **Magnet, rare-earth type (optional)**
» **Power supply, 9V** Maker Shed #MKSF3, $7
» **Ethernet cable**
» **Hookup wire**
» **Ethernet-to-wi-fi bridge (optional)**
» **Computer running Arduino IDE** install free from arduino.cc
» **Project code** download from makezine.com/36

TOOLS:
» **Soldering iron and solder**

Now you're in the device development page. Make a note of the Feed ID number at the top. Click on Add Channel to create 2 data streams: "openCount" for the number of fridge openings and "openDuration" for the duration of openings. Click on Save Channel. Your feed is ready to accept data.

You also need an API key to authorize your device to update the feed. Click on the Add Key button. Give the key a name, check all 4 permissions checkboxes, and click Save Key. You'll be shown a 48-character-long string of gibberish. Make a note of it. This key and the feed ID are what you'll need later.

2. Build the hardware.

This project requires very little soldering. The wiring diagram is shown in **Figure A**. The door sensor is a magnetic reed switch, cannibalized from a $2 door/window alarm. Pop open the case and solder a pair of wires across the

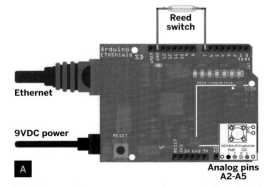

A

Reed switch

Ethernet

9VDC power

Analog pins
A2-A5

glass tube that is the switch (**Figure B**). Feed the wires through the power switch hole and snap the case back together.

Normally for switch inputs, you need a resistor to set the voltage when the switch isn't pressed. Instead, we'll use the Arduino's internal pull-up resistor on the switch pin. You'll see it enabled in the sketch like this:

B

```
pinMode(doorPin, INPUT);
digitalWrite(doorPin, HIGH);
```

Plug the Ethernet Shield into the Arduino, and the reed switch wires into the Gnd and Pin 7 positions on the shield. Finally, for a visual indicator, plug the BlinkM into pins A2–A5 (**Figure C**). BlinkMs speak the I²C serial protocol, which is normally shared with Arduino pins A4 and A5. We create a "virtual power supply" for the BlinkM by setting A3 high and A2 low.

C

3. Load the Arduino sketch.

Download the Arduino sketch, and libraries for Xively settings and BlinkM communication, from makezine.com/36. Open the sketch *CloudFridge0.ino* and click the *XivelyDetails.h* tab. Enter your Xively Feed ID and API key from Step 1 and save the file.

The sketch is based on the *PachubeClientString* example sketch in the Arduino Ethernet library, but uses DNS to connect to api.xively.com and uses DHCP to automatically get on your home network instead of requiring you to pick a static IP. It also uses a rollover-proof way to handling the **millis()** function which wraps around to zero after 49 days.

Every 100 milliseconds (ms), it reads the door sensor and updates the data. Every 30 seconds, it sends accumulated door data to Xively. To alter these intervals, change the values of the **doorUpdateMillis** and **postingInterval** variables, respectively.

The 2 data values being sent are the variables **doorOpenings** and **doorOpenMillis**. These values must first be turned into a string and that string's length determined. (This is because Xively uses **HTTP PUT** requests, which

require a **Content-length** header.) I used the **sprintf()** function:

```
char datastr[80];
doorOpenings=2;
doorUpdateMillis=13700;
sprintf(datastr, "%ld,%ld",
doorOpenings, doorOpenMillis);
Serial.println(datastr); // prints
"2,13700"
```

to turn the variables into a list of comma-separated values (CSV). Xively accepts CSV, XML, or JSON, but CSV is easiest for this application.

4. Test it out.

Place a magnet on the reed switch to trigger the "door closed" state. Connect an Ethernet cable to your Arduino and your home's router. Plug a USB cable between your Arduino and your computer and press Upload in the Arduino IDE. Open up the Arduino Serial Monitor and you'll see a status message about getting on the Ethernet and a test-send of data to Xively.

Remove the magnet, and the Arduino will print **doorOpen!** every 100ms until you put the magnet back. After a few seconds, it will make a successful **HTTP PUT** request of the door data to Xively. Your data is in the cloud.

CloudFridge

Feed ID 640418878
API Endpoint https://api.xively.com/v2/feeds/640418878

Channels Last updated 2 days ago ⟋ Graphs

openCount   0
Open

openDuration   0
msec

Create your 🐦 **Tweet**

Message (required)
The content of your new tweet.

Close the Fridge! Body Environment Feed

Body Triggering Datastream ID :
Body Triggering Datastream Value Value

F

6. Install it.

If you don't have an Ethernet hookup near your fridge, use any Ethernet-to-wi-fi bridge, such as an Airport Express or Linksys WET11.

I drew up an enclosure (thingiverse.com/thing:21939) with a lid that acts as a nice diffuser for the BlinkM LEDs, providing a glanceable status indicator. Off means it's just hanging out, blue means the door is open, and red means there was a network error.

Test-fit the reed switch and magnet to your fridge door gap with the door closed. The blue light should go out. (If the magnet seems weak, replace it with a rare-earth magnet.) Open your fridge and the blue light should go on (**Figure E**). Tape or hot-glue the switch in place. Your fridge is in the cloud!

5. View Xively data.

Xively's graphs give you a basic idea of what's going on (**Figure D**), but they're geared towards sampling. For accumulative data like our door openings, aggregation is better.

So it's more instructive when debugging to look at the data in XML format using query URLs. Xively has a great API (xively.com/dev/docs/api) that's very useful here. For every feed you create, there's an XML data URL. By adding query arguments to it, you can choose what time periods and data samplings you want to see. The most important argument is **interval=0**, which says to show all data. To list only data from the recent past, take the current time as shown by your XML feed, subtract a few minutes, and make that your **start=** argument, e.g. api.xively.com/v2/feeds/640418878.xml?interval=0&start=2013-07-22T23:30:00Z. The XML output lists each data stream, with an ISO 8601 time-stamp for every data point at 30-second intervals.

7. Send alerts.

Xively will send an **HTTP POST** to the URL of your choice when a specific condition is met. For instance, any time the fridge is open more than 20 seconds, I'm obviously just standing there searching for a snack. Let's send a tweet to publicly chastise me. Zapier is a service that lets you automate tasks between web apps (**Figure F**). Connect a Xively trigger to Zapier when **openDuration** exceeds **20000** and you can send a tweet via Twitter ("Close the fridge, Tod!"), activate a call or SMS via Twilio, or influence web apps. Learn how at xively.com/dev/tutorials/zapier/.

This is just the beginning. Apply this same setup to your front door, pet door, air conditioner, or any other device that wants to be on the Internet of Things. ◪

Tod E. Kurt and Mike Kuniavsky are the founders of ThingM (thingm.com), a ubiquitous computing/Internet of Things design studio and micro-manufacturer. Independently, they've written multiple books on embedded product design and hacking.

HOW TO BAKE AN ONION Pi

Written by Limor Fried and Phillip Torrone

Hack your Raspberry Pi into an anonymizing Tor proxy!

⚡ **TIME: 1–2 HOURS** ⚡ **COST: $90–$130**

Feel like someone is snooping on you? Browse the web anonymously anywhere you go with the Onion Pi Tor proxy. This is a cool weekend project that uses a Raspberry Pi mini computer, USB wi-fi adapter, and Ethernet cable to create a small, low-power, and portable privacy Pi.

Using it is easy-as-pie. First, plug the Ethernet cable into any internet connection in your home, work, hotel, or conference/event. Next, power up the Pi with the Micro-USB cable connected to your laptop, or with a wall adapter. The Pi will boot up and create a new secure wireless access point. Connecting to that access point will then automatically route any web browsing from your computer through the anonymizing Tor network. Your tracks are swept clean.

Nate Van Dyke

MATERIALS:

» **Raspberry Pi Starter Kit** #MSRPIK from Maker Shed, makershed.com. Our kit is the best way to get started using your Raspberry Pi. Includes Raspberry Pi Model B, 4GB SD Card, 5V 2A power supply, Micro-USB and HDMI cables, custom MAKE: Pi enclosure, Adafruit's Cobbler GPIO (General Purpose Input/Output) breakout, a breadboard for electronics prototyping, a selection of common components, and a copy of our bestselling book, *Getting Started with Raspberry Pi.*
» **Mini USB wi-fi module** Maker Shed #MKAD55

—OR—

» **Onion Pi Bundle (Tor Router) w/Mini Wi-Fi** Maker Shed #MSBUN44. For more experienced users who want to build a dedicated wireless Tor proxy. Includes Raspberry Pi Model B, Adafruit Pi case, Mini USB WiFi module, 10' Ethernet cable, Micro-USB cable, 5V 1A power supply, USB console cable, and 4GB SD card.

—OR—

» **Raspberry Pi Model B** Ethernet is required.
» **Raspberry Pi case** (optional)
» **Ethernet cable**
» **USB wi-fi adapter** that supports the RTL8192CU chipset
» **SD card** 4GB or more
» **5V Micro-USB power supply** rated at least 700mA

TOOLS:

» **Computer** Windows, Mac, or Linux
» **Router** with working internet connection
» **USB keyboard**
» **Display** with HDMI or composite video-in

What Is Tor?

Tor is an "onion routing" service: Internet traffic is wrapped in layers of encryption and sent through a random circuit of relays before reaching its destination. This makes it much harder for the server you're accessing (or anyone snooping on your internet use) to figure out who and where you are. It's an excellent way for people who are blocked from accessing websites to get around those restrictions. Journalists, activists, businesspeople, law enforcement agents, and even military intelligence operatives use Tor to protect their privacy and security online.

Why Use a Proxy?

You may have a guest or friend who wants to use Tor but doesn't have the ability or time to set it up on their computer. You may not want to, or may not be able to, install Tor on your work laptop or "loaner" computer. You may want to browse anonymously on a netbook, tablet, phone, or other mobile or console device that cannot run Tor and does not have an Ethernet connection. There are lots of reasons you may want to build and use an Onion Pi, not least of which is that it is an interesting way to learn about Raspberry Pi, network interfaces, and the Linux command line.

1. Prepare your SD card.

When you buy a Raspberry Pi, it may or may not come with an SD card. The SD card is important because this is where Raspberry Pi keeps its operating system and it's also where you'll store your documents and programs. Even if your Pi came with an SD card with the operating system already installed, it's a good idea to update it to the latest version, as improvements and bug fixes are going in all the time.

Anonymously Yours
WARNING:

Before you start using your proxy, remember that there are a lot of ways to identify you, even if your IP address is "randomized." So delete and block your browser cache, history, and cookies — some browsers even allow "anonymous sessions." Do not log into existing accounts with personally identifying information (unless you're sure that's what you want to do). Use SSL whenever available to encrypt your communication end-to-end. And visit torproject.org for more info on how to use Tor in a smart, safe way.

This tutorial is a great way to make something fun and useful with your Raspberry Pi, but we can't guarantee it's 100% anonymous and secure. Be smart and paranoid about your Tor usage.

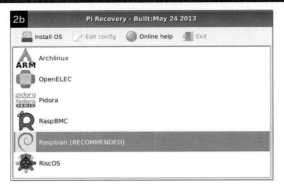

NOTE : This tutorial assumes you'll be using Raspbian, and may not work exactly as written with other Linux distributions.

Experienced users have many options for preparing an SD card. We recommend new users visit raspberrypi.org and follow the instructions in their Quick Start Guide for formatting an SD card and installing the official New Out Of Box Software (NOOBS) package. Briefly, the steps are:

1a. Format the card. The Raspberry Pi Foundation recommends using the SD card foundation's official formatting tool, SD Formatter, which is available for Windows, Mac, and Linux. The settings may vary depending on your OS. Refer to the Quick Start Guide for details.

1b. Download NOOBS. You can get the .ZIP archive directly from the Raspberry Pi website, one of several mirror servers, or through BitTorrent.

1c. Extract the NOOBS archive to your SD card. The contents of the archive, including the file *bootcode.bin* and the *images* and *slides* folders, should be in the top-level directory.

2. Boot and configure.

If you want to mount your Pi in a case, now's a good time.

2a. Insert the SD card you just prepared into the Pi's drive slot, being careful to note the correct orientation of the contacts. Connect your display and keyboard before plugging in the Micro-USB power cable. The Pi should boot automatically.

2b. Install Raspbian. From the NOOBS boot screen, select Raspbian, press Enter, and confirm that you want to overwrite the disk. When installation is complete, press Enter again to dismiss the notice, and your Pi should reboot automatically.

2c. After a lot of scrolling text, you'll arrive at the `raspi-config` options screen. Using the arrow keys to navigate and Enter to select, first update the default password ("raspberry") for the default user account ("pi") to a secure phrase known only to you.

TIP : You may notice a short lag between selecting options or entering commands and the system's response. This is normal. Be patient.

2d. Select Internationalisation Options and set the time zone, language, and keyboard layout options to match your preference. Then select Finish and press Enter.

3. Connect Ethernet/Wi-Fi.

For most home networks, you should also be able to connect to the internet through the Ethernet connection via your router without any further configuration. After `raspi-config` exits,

```
┤ Raspberry Pi Software Configuration Tool (raspi-config) ├
 Setup Options

      1 Expand Filesystem          Ensures that all of the SD card storage is available to the OS
      2 Change User Password       Change password for the default user (pi)
      3 Enable Boot to Desktop     Choose whether to boot into a desktop environment or the command-line
      4 Internationalisation Options Set up language and regional settings to match your location
      5 Enable Camera              Enable this Pi to work with the Raspberry Pi Camera
      6 Add to Rastrack            Add this Pi to the online Raspberry Pi Map (Rastrack)
      7 Overclock                  Configure overclocking for your Pi
      8 Advanced Options           Configure advanced settings
      9 About raspi-config         Information about this configuration tool


                   <Select>                                              <Finish>
```

3a

MAKE ME A SANDWICH.

SUDO MAKE ME A SANDWICH.

WHAT? MAKE IT YOURSELF.

OKAY.

Classic *xkcd* webcomic #149, "Sandwich."

Randall Munroe

you'll be presented with the Raspbian command prompt:

`pi@raspberrypi ~ $_`

3a. When you see the prompt, connect your Pi to your router using a standard network cable. As soon as you plug your Pi in, you should see its network LEDs start to flicker.

3b. At the Raspbian command line, type in:
`sudo wget makezine.com/go/onionpi`
The Linux command `sudo` allows one user to assume the security privileges of another, commonly the superuser or root. (Think: " superuser do.") The next command, `wget`, will not run correctly unless preceded by `sudo`.

NOTE : Linux user rights and privileges can get pretty complicated, but as a general rule, you'll need to `sudo` any commands that involve making changes to the disk. Read-only commands, like listing directories or displaying (without modifying) the contents of files, can usually be executed without `sudo`.

The command `wget` instructs the operating system to retrieve a file from the web, and takes as argument the web address of the file to be retrieved. In this case, we're grabbing a pair of *shell scripts* that will automate much of the fiddly typing for configuring your Pi as a wireless access point.

TIP : If you get tired of typing `sudo` all the time, the command `sudo su` allows you to become the superuser as long as you want.

When you understand what the command is supposed to do, press Enter to execute it. If your Ethernet connection is working, you'll shortly be notified that the file has been saved.

If your Ethernet connection is *not* working, you'll see an error message (such as `failed: Name or service not known`). Make sure that your Pi is correctly connected to your router, the network cable is good, and your router is correctly configured for DHCP (Dynamic Host Configuration Protocol).

3c. Don't plug in your wi-fi adapter yet — you'll crash the Pi and corrupt the SD card. First, turn off your Pi by entering `sudo halt`. After shutdown, plug in the wi-fi adapter. Now restart your Pi by cycling the power.

4. Set up the "PiFi" access point.

Now we'll set up the Pi to broadcast a wi-fi service and route wireless internet traffic through the Ethernet cable. One of the great things about Linux is that every little detail of a system's configuration can be easily modified to suit your application by typing in commands or modifying the contents of text files.

The tradeoff is that the details can get pretty complicated, and you have to know what you're doing to understand exactly what needs to be changed, and how.

To make the process easier, we've prepared a script (which you just downloaded with `wget`) that will automatically make these changes for you (**Figure 4**). If you just want to get it working, all you have to do is run the script, as explained below.

4a. After your Pi reboots, you'll be prompted to log in. Enter the default user ID "pi" followed by the password you set from `raspi-config`.

4b. At the Raspbian command prompt, enter these commands to extract the shell scripts:
```
sudo unzip onionpi
sudo bash pifi.sh
```

We just made friends with `sudo`; now it's time to meet `bash`, the Linux *command-line interpreter*. In fact, you've already been introduced: Whenever you enter text at the command prompt, you are interacting with `bash`, which is the program that processes what you've typed and figures out what to do with it. `bash` runs automatically whenever you're working from the Linux command line, but can also be called as a command, itself, to execute a script file.

In this case, we're telling `bash` to read through the script `pifi.sh` and execute each line of text as if it had been typed in at the command prompt.

4c. Press Enter and you'll soon see the script splash screen, with the option to start the script or abort. Press Enter again to start.

4d. When prompted, enter the name (SSID) for your new wireless network, and the password required to access it.

When the script is complete, your Pi should reboot automatically, after which you should be able to detect your new "PiFi" network from nearby computers, smartphones, and other wi-fi appliances. Log on to the wireless network using the password you just set, open a web browser, and navigate to your favorite web page to verify that everything is working properly.

If you just want to configure your Pi as a wireless access point, you're done! You shouldn't even have to log in to Raspbian

NOTE: Both network name and password can be updated later by editing the config file with any text editor.

```
4   pi@raspberrypi ~ $ sudo bash pifi.sh


              .~.     .~.
          /  '.\ : /.'  \
        /   / .~'~.  \   \
      |   |  /  : :  \  |   |
      |   |  ~ (_) (_) ~  |   |
      |   | ( : ~.~.~: ) |   |
      |   |  ~.~ (_) ~.~  |   |
       \   \   ( :~: )   /   /
         \  '._ '~_~' _.'  /
              '~.~'


           Raspberry PiFi

This script will configure your Raspberry Pi as a wireless access point.
Press [Enter] to begin, [Ctrl-C] to abort...
```

NOTE: For a slower and more instructive experience, we recommend opening the *pifi.sh* script (which is just a text file) in another computer and typing in the commands by hand, to get a feel for what each one does and how the system responds. The script file also contains comments that explain each step in more technical detail, for those who are interested.

again; the Pi will now automatically function as a wireless router whenever it's on.

5. Install tor.

To continue setting up your Pi to anonymize your wi-fi traffic with Tor, log in to Linux again and run the second script with:
`sudo bash tor.sh` **(Figure 5)**

This script is less complicated. Basically, it installs and configures the Tor software, then updates your IP tables to route everything through it. As always, it's a good idea to read through the commands and comments in the script file before running it. More technical detail is available there.

The Pi will automatically reboot again when the script is done. Your Tor proxy may not work until the reboot is complete.

6. Browse anonymously.

When your Pi has finished rebooting, log on to your "PiFi" wireless network from a nearby computer, smartphone, or other wi-fi appliance. Then open your favorite internet browser and visit check.torproject.org. If your Onion Pi is working correctly, you should see something like **Figure 6**.

Going Further

We use Ethernet because it requires no configuration or passwords — just click the cable to get DHCP. But if you want, it's not too hard to set up a wi-fi-to-wi-fi proxy. You'll need to use two wi-fi adapters and edit the settings in `/etc/networks/interfaces` to add the `wlan1` interface with SSID and password to match your internet provider. See makezine.com/go/pifi2wifi for more details.

It's also pretty easy to configure Tor to give you a presence in any country you choose. For example, here's a *torrc* configuration file that sets up a Pi at IP address 192.168.0.178 to appear "present" in Great Britain:

```
Log notice file /var/log/tor/notices.log
SocksListenAddress 192.168.0.178
ExitNodes {GB}
StrictNodes 1
```

This script will auto-setup an Onion Pi Tor proxy for you.
Press [Enter] to begin, [Ctrl-C] to abort...

You'll also need to configure your browser to use a SOCKS5 proxy on 192.168.0.178 (or whatever your Pi's IP address may be), port 9050.

If you like using Tor, you can help make it faster by joining as a relay, or increase its effectiveness by becoming an exit node. Check out torproject.org for details.

Finally, if you want to support Tor but can't run your own relay or exit node, please consider donating to the project to help cover development, equipment, and other expenses. Your donation is even tax-deductible if you live in the United States. ◪

Limor Fried is owner of Adafruit Industries, an open-source hardware and electronics kit company based in New York City.

Phillip Torrone is an editor at large of MAKE magazine and creative director at Adafruit.

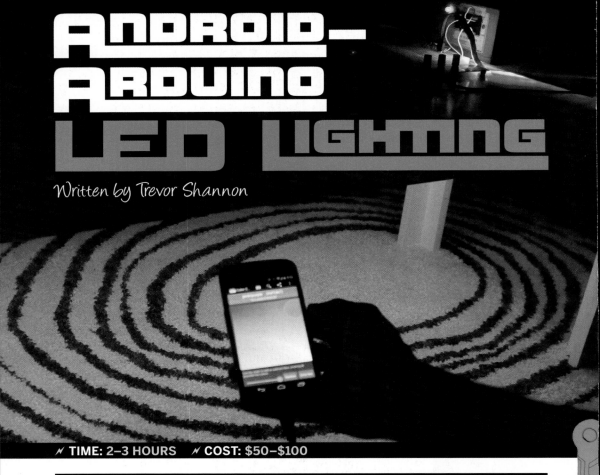

ANDROID– ARDUINO LED LIGHTING

Written by Trevor Shannon

⚡ TIME: 2–3 HOURS   ⚡ COST: $50–$100

Connect a smartphone to a microcontroller and make LEDs glow any color with a finger swipe.

Want a touch-controlled light show? Attach this color-changing RGB LED strip to the underside of a coffee table, your bike, or anything else that needs a bit more color, and use your phone to run the display.

There are many ways to connect a smartphone to an embedded microcontroller like an Arduino. In this project you'll use an Android device in USB host mode. This way, the phone both powers and communicates with the Arduino. Even though the Arduino is connected via USB, the communication happens via serial, just like when it's connected to your computer.

Once your phone can talk to an Arduino, a whole world of new projects opens up!

1. Build the circuit.

RGB LED strips usually have 4 wires: one for power, and one each for red, green, and blue control. When the strip is powered, pulling any of the control lines to ground will cause that color LED to light up at full brightness. Using pulse-width modulation (PWM) on those control lines allows you to modulate the brightness of the lights.

A strip 1m long can draw nearly 1A when the red, green, and blue LEDs are on full

MATERIALS:

» **Arduino Uno microcontroller board** Maker Shed #MKSP11 (makershed.com), RadioShack #276-128, or other USB-to-serial Arduino
» **Android smartphone or tablet capable of USB host mode** See makezine.com/go/android-arduino for a partial list of compatible phones.
» **USB cable, On-The-Go (OTG), micro-B male to standard-A female** Buy one for $5, or make your own (see Step 2).
» **RGB LED strip, non-addressable, 1m (or longer)** such as Adafruit Industries #346
» **Transistors, NPN, high-power** such as TIP31, RadioShack #276-2017
» **Resistors, 1kΩ (3)** RadioShack #271-1321
» **AC adapter, 12V, 1A** RadioShack #273-316 or 273-462
» **Breadboard** RadioShack #276-098, Maker Shed #MKKN3 or MKEL3
» **Jumper wires** RadioShack #276-173, Maker Shed #MKSEEED3

TOOLS:

» **Soldering iron and solder**
» **Electrical tape**
» **Wire strippers**
» **Razor blade**
» **Computer running Arduino IDE software** free from arduino.cc
» **Project code** download from makezine.com/36

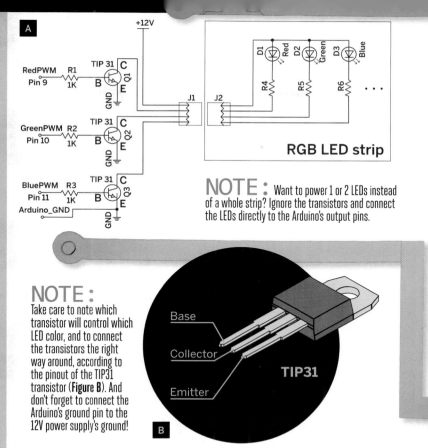

A

RedPWM Pin 9 — R1 1K — B — TIP 31 C Q1 E — GND

GreenPWM Pin 10 — R2 1K — B — TIP 31 C Q2 E — GND

BluePWM Pin 11 — R3 1K — B — TIP 31 C Q3 E — GND

Arduino_GND

+12V

D1 Red   D2 Green   D3 Blue

R4   R5   R6   . . .

J1   J2

RGB LED strip

NOTE: Want to power 1 or 2 LEDs instead of a whole strip? Ignore the transistors and connect the LEDs directly to the Arduino's output pins.

NOTE: Take care to note which transistor will control which LED color, and to connect the transistors the right way around, according to the pinout of the TIP31 transistor (**Figure B**). And don't forget to connect the Arduino's ground pin to the 12V power supply's ground!

Base
Collector
Emitter
TIP31

B

brightness. An Arduino's output pins can only supply around 40mA each, so you'll have to help it out with a driver circuit to boost the power. This circuit takes 3 PWM signals from the Arduino and uses them to drive 3 transistors supplying power to the red, green, and blue LEDs — giving you full control over the brightness of each color, so you can mix them to create any color.

The driver circuit (**Figure A**) is a basic transistor amplifier repeated 3 times. A 5V, low-current PWM signal comes from the Arduino to the transistor's base (B) via a 1K resistor. This signal switches the transistor, allowing it to conduct higher current at 12V across the collector (C) and emitter (E) and through the LEDs (**Figure B**). The transistor can switch fast enough that the LED power is PWM'ing just like the input signal, resulting in the desired brightness control.

Your completed circuit should look like **Figure C** on the following page. The outputs of the circuit go to a 4-pin header on the right, ready to accept an RGB LED strip.

2. Hack a cable (optional).

Many Android phones and tablets are USB On-The-Go (OTG) devices, which means they can act as the USB host (providing

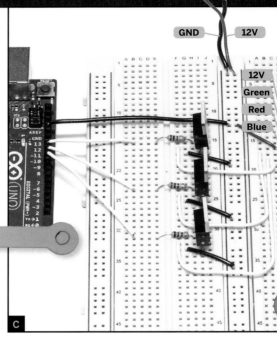

power) or slave (receiving power). In order for your phone to host a USB device such as an Arduino, you need a USB OTG cable. They only cost a few bucks, but it's more fun to make one from an old USB cable. Follow the instructions at makezine.com/projects/usb-otg-cable.

3. Install the Android app.

By adding a few lines of code to an Android app, you can send any kind of data to the

Arduino via your USB OTG cable. In this case, you'll send brightness values between 0 and 255 for the red, green, and blue LEDs. The critical piece of the puzzle here comes from Mike Wakerly. He wrote a fantastic driver for the USB-to-serial chips used in Arduino boards, called *usb-serial-for-android*. Sending data to an Arduino from your phone is as simple as calling `device.write()` within your Android app!

One great example of this is Katie Dektar's open-source app called Color Namer (github. com/kaytdek/ColorNamer). I simplified it into a version called Arduino Color (**Figure D**). Download the app and installer with your Android device at trevorshp.com/creations/ArduinoColor.apk. (You can also

grab the complete source code at github.com/trevorshannon/ArduinoColor.)

4. Code your own app (optional).

If you're developing your own app, download Wakerly's driver from code.google.com/p/usb-serial-for-android, and copy the JAR file into the *[your_app_root]/libs* directory. This is a pre-compiled library that includes all the necessary functions to open, close, write to, and read from a serial device from your Android app.

Next, copy the *device_filter.xml* file into the *[your_app_root]/res/xml* directory. This file acts as a lookup table for USB devices connected to your phone. When you connect an Arduino, its device ID (set by the manufacturer of the USB-to-serial chip) is sent to the phone. If this ID appears in your app's *device_filter.xml* file, then your app will automatically open. (The Arduino's ID is already in Wakerly's *device_filter.xml* file, so you shouldn't need to modify that file.) Finally, make your app aware that it should be looking for USB devices in the first place. Open your app's *AndroidManifest.xml* file and add the appropriate `<intent-filter>` information within your `<activity>` tags:

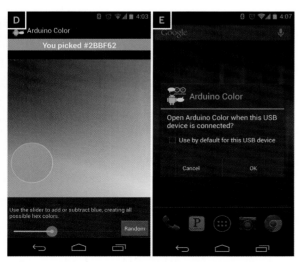

```
<intent-filter>
<action android:name="android.
hardware.usb.action.USB_DEVICE_
ATTACHED" />
  <action android:name="android.intent.
action.MAIN" />
  <category android:name="android.
intent.category.LAUNCHER" />
</intent-filter>
<meta-data
android:name="android.hardware.usb.
action.USB_DEVICE_ATTACHED"
  android:resource="@xml/device_filter"
/>
```

To integrate Wakerly's library, you also need code that opens the serial device when the app begins and closes it when the app stops. Then you can send data to the opened serial device using the write() function. For example, the Arduino Color app will send color data via serial when the user touches the color picker. You can find code examples from Arduino Color at makezine.com/36.

5. Upload the Arduino firmware.

The Arduino code for this project is fairly simple; download the *android_leds.ino* file from makezine.com/36, then open it in the Arduino development environment (IDE), and upload it to your Arduino board.

The code uses the Serial library, which is provided with the Arduino IDE and makes it easy to read data sent via serial to the microcontroller. The code endlessly loops while querying the number of available bytes on the serial port. When there are at least 3 bytes available, the bytes are read and stored in an array as red, green, and blue brightnesses. Those brightness levels are then used to set the PWM outputs accordingly.

Notice that there are constants called max_red, max_green, and max_blue. In theory, if you turn on the red, green, and blue LEDs at maximum brightness, the resulting light should be white. In practice, your LED strip may have a slight color cast (generally green or blue). You can compensate for this by

tweaking the max value constants that are called by the map function.

6. Connect it all together.

Connect the Arduino to your phone and a notification will pop up asking if you'd like to open the Arduino Color app (**Figure E**). Yes!

Connect the LED strip to your driver circuit, and turn on the 12V power supply. Using the app, you should now be able to adjust the color of your light strip in real time (**Figure F**).

Attach the light strip to the underside of a coffee table (page 80), bring it to a nighttime party, or stick it on your bike!

Other Explorations

Now that you know how to talk to an Arduino with an Android app, take this project further.
» Go wireless: Add a Bluetooth or wi-fi module to your Arduino and get rid of the cable.
» Make lights that change color based on the orientation of your phone.
» Use the GPS in your phone as well as some fancy sensors connected to the Arduino to make a location-aware data logger.
» Use an individually addressable LED strip and adjust the code for more control over the light show. ▰

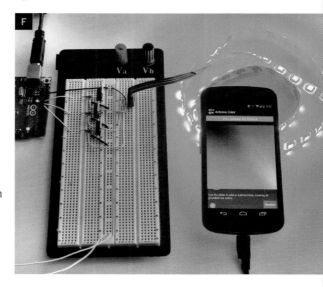

Trevor Shannon (trevorshp.com) likes experimenting, learning, and building cool stuff. If he gets really lucky, the projects he builds actually work!

SKILL BUILDER

MAKE YOUR OWN D*MN BOARD

Use EAGLE to design a bare bones Arduino board.

By **Shawn Wallace**

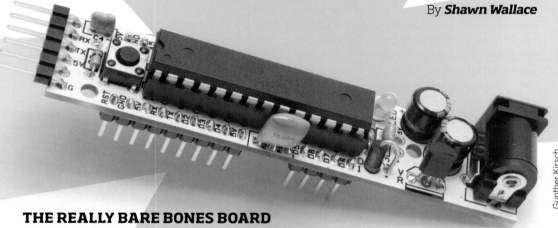

Gunther Kirsch

THE REALLY BARE BONES BOARD

(RBBB) and its schematic were originally designed by Paul Badger of Modern Device. It's pretty much the bare minimum needed for a useful Arduino-type development board. And once you build the schematic in EAGLE, it's easy to incorporate in other designs.

Using EAGLE

EAGLE is a collection of programs, each serving one part of the design process. We'll focus on Schematic Editor and Board Editor; other modules include Autorouter (trace layout AI), Parts Editor, CAM Processor (for creating machining-ready files), and a scripting interface for writing ULPs (User Language Programs).

Components are chosen from the parts library. Parts always have two representations: one for the schematic and one for the board. Most components come in different physical "packages," and EAGLE's libraries allow multiple packages to be associated with the same schematic symbol. To simplify things, I've created a custom RBBB library that includes only the parts in the packages you'll need. Once you're comfortable, SparkFun's library is a good place to start expanding your horizons.

EAGLE separates CAD drawings into color-

TIP: Check the toolbar often to make sure you're drawing into the right layer.

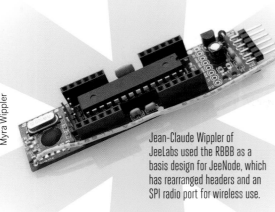

Jean-Claude Wippler of JeeLabs used the RBBB as a basis design for JeeNode, which has rearranged headers and an SPI radio port for wireless use.

TOOLS & MATERIALS

There are many tools for creating printed circuit boards, from the open source KiCAD to various online services. CadSoft's EAGLE is a favorite in the open hardware community. Here you'll learn the basics of PCB design in EAGLE, plus how to build the essential core of an ATmega microcontroller board.

THE RBBB COMPONENTS

» **Microcontroller IC, ATmega328P**
» **IC socket, 28-pin (optional)**
» **Electrolytic capacitors, 47µF (2)**
» **Ceramic capacitors, 0.1µF (2)**
» **Resistor, 10K ⅛W**
» **Resistor, 1K ⅛W**
» **Ceramic resonator, 16MHz**
» **Switch, momentary pushbutton**
» **Voltage regulator, low-dropout (LDO), L4931**
» **Diode, 4005**
» **Power jack**
» **Pin header, 1×6**

✳ ✳ ✳ ✳ ✳ ✳ ✳ ✳ ✳ ✳ ✳ ✳ ✳ ✳ ✳ ✳

coded functional layers. In Schematic Editor, part outlines are on the Symbols layer, labels on the Names/Values layers, and nets (aka signals) on the Net layer. You can show or hide the layers using the View menu.

PCB design takes several stages. This article guides you as far as drawing and validating the schematic, which should be enough to whet your appetite. When you're ready, the remaining stages are covered in online tutorials at makezine.com/36.

1. Getting started.

Download EAGLE Light Edition from cadsoft usa.com. Also grab the RBBB library from moderndevice.com/downloads and place it in EAGLE's /lbr directory.

1.1. Open Schematic Editor and select File → New → Schematic. You'll get a warning that

EAGLE DESIGN WORKFLOW

I. **Design and Sourcing:** Find parts, read datasheets, download or draw footprints.

II. **Schematic Layout:** Connect parts electrically with signals.

III. **Electrical Rule Check (ERC):** Run an AI helper to identify schematic errors.

IV. **Board Layout:** Place parts on the board and draw the actual traces connecting them.

V. **Design Rule Check (DRC):** Run an AI helper to identify layout errors.

VI. **Generate Gerber and drill files:** Run CAM Processor to generate machining files.

"no forward/backward annotation will be performed." Normally you'll have both Schematic and Board Editor open at once so EAGLE can keep them synced; this message just says that there's no board file open yet.

1.2. Choose the Add tool, grab the Frame part from the RBBB library, and place it on the canvas. It's not essential, but it's handy so you don't have to keep resizing your window as the schematic grows.

2. The power circuit.

These 5 parts accept DC input power and output a clean 5V supply.

The centerpiece is the voltage regulator. The key specs for a regulator are output voltage, maximum output current, and dropout voltage, which is the minimum difference needed between input and output. The L4931 has a dropout of just 0.4V, so it can reliably output 5V given 6V (e.g. from 4 AA batteries). It provides current up to 250mA, which is plenty for most microcontroller applications.

The 2 electrolytic capacitors filter noise on the input supply, pick up slack when batteries fade, and handle momentary power demand spikes. The regulator's datasheet recommends 2.2µF filter caps minimum; we'll use 47µF. That should be stiff enough for even the noisiest battery-powered breadboard experiments.

The power jack may seem big, but this is one of those human interface decisions that really matter — this jack easily accommodates a "wall wart" power supply. We'll position it on the board where it can be snipped off if it's

NOTE: Our diagrams may look different from your screen, at first. Remember to use Smash to separate the labels, then use Move so they're readable.

RULES OF THUMB

» A dot indicates that intersecting signal lines are connected; lines that cross without a dot are just overlapping.

» Stay on the grid. The most common mistakes come from parts not lining up, looking like they're connected when they aren't. Set it to 0.1 (100 mil) to start.

» Each line (or net, or signal) has a unique name, and all signals with the same name are automatically connected, even if there's not a line between them. This is handy if you have so many parts that drawing every line would make a pile of spaghetti.

» Each part should have a visible name and a value. You'll need to move the labels around to make them readable; use the Smash tool to separate labels from parts, then Move and Rotate to position them.

» Drawing a schematic is about communicating a circuit assembly to someone else. Think about whether you're providing everything they will need.

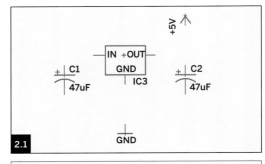

2.1

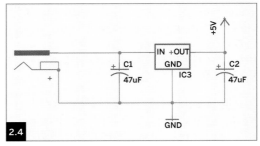

2.2

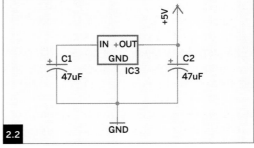

2.4

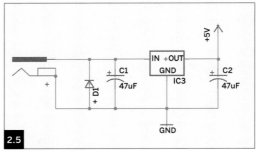

2.5

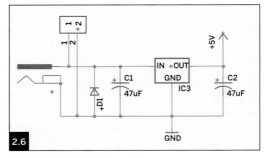

2.6

unneeded, and provide a 2-pin auxiliary power header for optional wiring.

Almost any power diode will serve for short-circuit protection; this one is a common 4005 rated 1A, wired in parallel to prevent damage if power is accidentally connected backwards. Some designs opt for a series protection diode, but that would introduce a 0.7V drop we can't afford if we want to run from 6V.

2.1. Select the Add tool, choose RBBB → Regulator, and place the regulator somewhere in the upper left quadrant. Use the Add tool to place the 2 Electrolytic_Caps, plus GND and +5V supply signals (under RBBB → Supply), as shown.

2.2. Use the Net tool to connect the caps' positive sides to the regulator's input and output, and their negative sides to the regulator's ground lead. Connect the regulator's ground lead to GND, and its output to +5V.

2.3. Use the Value tool to assign each capacitor a value of 47µF.

2.4. Add RBBB → Power_Jack beside the regulator input. This jack is center-positive, which (outside of musical electronics) is pretty standard. Connect its center pin to the regulator input and its sleeve to ground. If you get a "Connect Net Segments?" dialog, select Yes.

2.5. Add the diode. It will appear horizontally. Use the Rotate tool to turn it cathode/negative side up. Then use the Net tool to connect it across the regulator input and ground.

2.6. Finally, drop in the 1×2 header (RBBB → 1x2_Pinhead) for the optional power input. Position it with Rotate, then connect one pin to power and one to ground.

3. The microcontroller and headers.

The headers connect to the microcontroller's general-purpose inputs and outputs, and they provide "pans" for soldering wires or pins.

3.1. Add the microcontroller chip (RBBB → ATmega) somewhere in the middle of the schematic, and connect its 2 ground pins to the GND supply signal.

3.2. Place a 0.1µF capacitor near the ATmega's

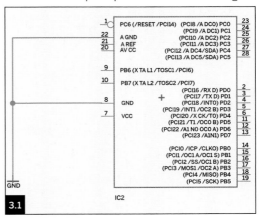

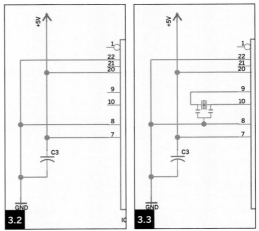

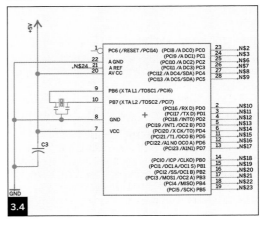

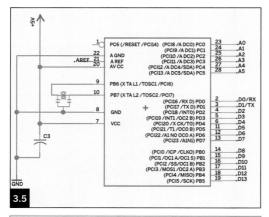

3.5

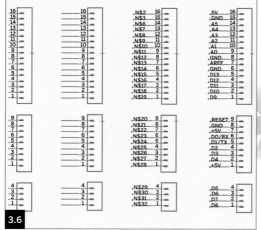

3.6

3.5. You'll see EAGLE gives a default label to each signal, something like N$2. Now use the Name tool to rename each signal to match its ATmega pin name. This is a bit tedious; if anyone knows a better way than manually, please let me know!

3.6. Add the 1×16, 1×9, and 1×4 header blocks and repeat the process of adding signals, adding labels, and changing names.

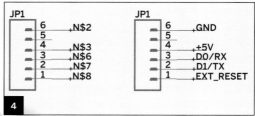

4

4. The serial header.

Our data interface is a 6-pin header connected to the ATmega's serial port. The original Arduino had a USB port and an FTDI chip to translate from USB to TTL UART protocol. Moving this chip off the board and into the cable is an easy way to pare down the design. FTDI sells a cable with a 6-pin connector, or you can use a USB-BUB or FTDI Friend.

NOTE: "Receive" on the board connects to "transmit" on the cable.

Place the 6-pin header and connect the RX (receive), TX (transmit), GND, +5V, and EXT_RESET pins.

power supply pin (7), and connect it to power, ground, and pin 7.

3.3. Place the resonator near clock pins 9 and 10, and connect its 3 pins as shown. Make sure the center pin goes to ground.

3.4. Create a signal line for all the other pins. Don't forget analog reference (AREF) on the left. Use Net to connect a short signal to each pin, then use Label (right underneath) to label each one.

NOTE: EAGLE's "group move" function does not work like most modern drawing software. If you need to move several parts at once, first use Zoom to zoom out, then choose Select and click-drag around the parts to be moved. Choose Move, left-click on the selected objects, Ctrl-right-click to get a menu, and choose "Move Group."

5. Amenities.

Many designs skimp on human factors, with tiny buttons, inconvenient part arrangements, and indecipherable labels. Don't do this. Our design includes a generously sized reset button and an LED "pilot light" to clearly show when the board is powered.

5.1. Position the switch, the 10K resistor, and the remaining 0.1µF capacitor as shown. Connect one side of the switch to ground. The switch is normally open, but pulls RESET

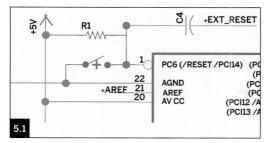

5.1

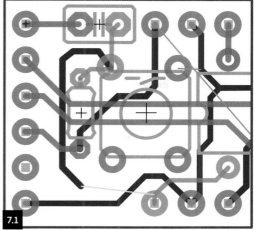

7.1

low when pressed. The 10K resistor is connected between RESET and +5V so that the pin is not "floating" (which can cause erratic behavior) when the switch is open. During pro- gramming, the host computer pulls EXT_RESET low to reset the chip before the bootloader starts loading code. The small capacitor keeps the timing of this pulse within the limits expected by the system.

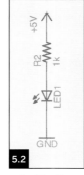

5.2

5.2. Finally, add an LED in series with a 1K current-limiting resistor.

6. Check the circuit.
Run the Electrical Rule Check by selecting ERC from the Tools menu. You'll get a list of errors and warnings; click on them to make any needed fixes. A common mistake is for lines to be very close without actually connecting.

Going further
Once you've passed the ERC, select File→ Switch To Board and you'll be asked if you want to create a new board from the current schematic using the Layout Editor. Yes! From now on you'll want to have both schematic and board windows open whenever you're work- ing. EAGLE will keep the two in sync, but only if they're both open.

Now you'll route the actual copper traces on your board. EAGLE can do this automatically, but it's a good idea to try your hand at manual routing, just to understand it. First you'll draw the boundary of the board and set the spacing grid. Then you'll lay out the power circuit, place the components, and route the GPIO traces, connecting top to bottom traces where neces- sary (**Figure 7.1**). Finally, you'll add a ground plane to the bottom of the board (**Figure 7.2**).

After routing, the only steps left are to create the silk-screen layer and the Gerber and drill files to be sent to the PCB house for manufac- turing. Follow our step-by-step instructions for laying out your board and outputting these PCB files, in parts 2 and 3 of this tutorial at makezine.com/36. ◪

Shawn Wallace is a MAKE contributor, artist, programmer, and editor living in Providence, R.I. He designs open hardware kits at Modern Device and organized the Fab Academy at the Providence Fab Lab. He makes iPhone synthesizers with the Fluxama collective and is a member of the SMT Computing Society.

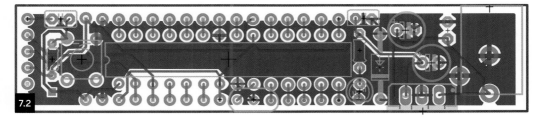

7.2

Nuclear **Fusor**

Written by *Dan Spangler*

⚡ **TIME:** 2–4 DAYS   ⚡ **COST:** $200–$500

Build yourself a star in a jar.

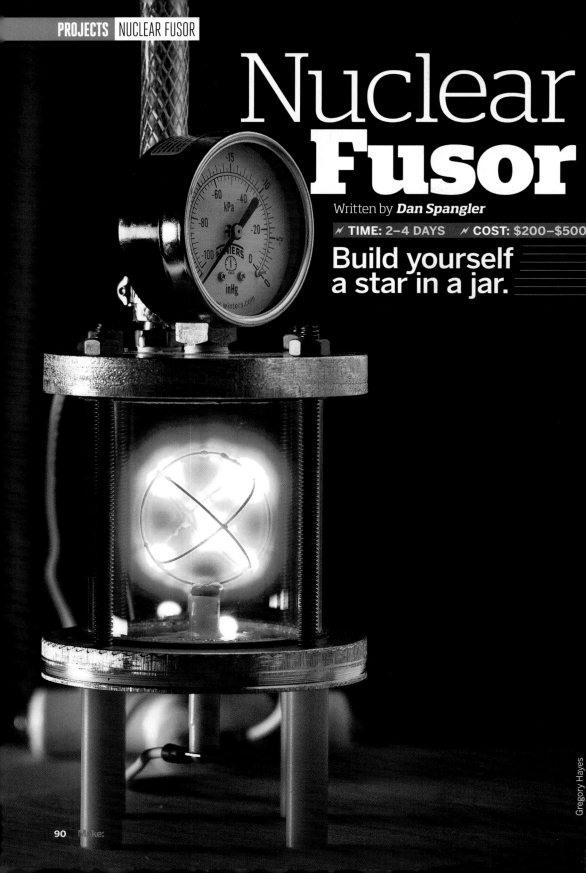

Gregory Hayes

Get Your Hot Fusion Here

➕ Nuclear fusion is the process of squeezing two atoms together so tightly that their nuclei fuse, creating a heavier atom and releasing a blast of energy. Fusion creates the inferno inside the sun – and the hydrogen bomb – but no one has yet harnessed its enormous power for peaceful uses.

They've tried, however, often with skepticism. In 1989, physicists Martin Fleischmann and Stanley Pons announced they'd achieved "cold fusion" of hydrogen into helium at room temperature, only to face withering scorn when others failed to replicate their results.

Luckily, DIY nuclear engineers can achieve honest-to-goodness "hot fusion" right at home by making a Farnsworth-Hirsch fusion reactor, or *fusor* for short.

This mini fusor is a demonstration version – while it generates only insignificant quantities of fusion products, it does show how inertial electrostatic confinement (IEC) reactors use kinetic energy to cause fusion. It's also a good introduction to high-voltage power supplies and vacuum systems. The skills the project imparts will help you tackle bigger fusors and other projects involving plasma and high-energy physics.

Plus, the fusor just looks totally cool. An eerie purple-blue glow emanates from the reactor, and a really well-made fusor can produce a mesmerizing phenomenon called a "star in a jar."

Curious? Read on…

WARNING:
This project uses high voltages at potentially lethal currents. A high-vacuum apparatus may implode if improperly handled. This device may produce ultraviolet and x-ray radiation. Do not attempt to build or operate it unless you are capable of safely using high voltage and vacuum equipment.

MATERIALS

- » **Neon sign transformer, 12,000V (12kV), without ground fault protection** Check eBay or your local neon shop.
- » **Variable transformer, 110V–140V, 5A** aka "Variac"
- » **Vacuum pump, 2 stage, 0.025mm Hg (25 micron) minimum vacuum rating** The higher the cubic-feet-per-minute (CFM) rating, the better.
- » **Diodes, high voltage, 0.1A 20kV (2)** Search hvstuff.com and buy extra in case you accidentally fry your rectifier.
- » **Vacuum gauge with ¼" NPT male fitting** such as Amazon #B0087UD1GA
- » **PVC tubing, flexible braided, ⅜" ID, 2' length** Braided reinforcement keeps it from collapsing under high vacuum.
- » **Barbed hose fittings, brass, ⅜" hose ID × ¼" NPTF male pipe thread (2)**
- » **Additional brass pipe fittings (optional)** if needed to connect your vacuum pump to the barbed hose fittings
- » **Hose clamps, for ⅜" tubing (2)**
- » **Spade terminals, ¼" wide, for 16–14 AWG wire (8-11)**
- » **Ring terminals, ¼" wide, for 16–14 AWG wire (3)**
- » **Oil cup cylinder, borosilicate glass, 3" OD × 2¹⁹⁄₃₂" ID × 3" high** McMaster-Carr #1176K27, mcmaster.com
- » **Nylon spacers, unthreaded, ½" OD, 2" long, for ¼" screws (4)** McMaster-Carr #94638A29 or 94639A089
- » **Aluminum bar stock, alloy 6061, ½"×5"×12"** McMaster-Carr #8975K436
- » **Threaded rods, plain steel, ¼-20 × 7" (4)** McMaster-Carr #91565A566
- » **Standoff, round, aluminum, female threaded, ¼" OD, 2" long, for #10-32 screws** McMaster-Carr #93330A493
- » **Ceramic tube, nonporous, high-alumina, 0.375" OD, ¼" ID, 12" long** McMaster-Carr #8746K18
- » **High-voltage wire, 302°F, 20 AWG, 0.138" OD, 20,000VDC, 6'** McMaster-Carr #8296K15
- » **Dow Corning High-Vacuum Grease** McMaster-Carr #2966K52
- » **Wire, stainless steel (type 302/304), soft temper, 0.032" diameter, 4'** McMaster-Carr #8860K14
- » **Hex nuts, ¼-20 (4)**
- » **T-nuts, ¼-20 (4)**
- » **Machine screws, stainless steel, #10-32 × ½" (5)**
- » **Hex nuts, #10-32 (3)**
- » **Plywood, ½" nominal, 8"×8"**
- » **Self-adhesive rubber feet (4)**
- » **Rubber, ¹⁄₁₆"-thick sheet**
- » **PVC pipe, ½" nominal, 6" length**
- » **PVC pipe fittings, ½" nominal, slip fit: end caps (3) and tee (1)**
- » **Insulated wire, stranded, 16 AWG**

(Continued on following page)

» Mineral oil, 16 fl oz
» Rubbing alcohol
» AC power cord, 3-wire, 4' (optional)
» Male plug with ground prong (optional)

TOOLS

» **Drill press and drill bits**
» **Hole saw, bimetal, 4¼" diameter, 1½" depth, ⁷⁄₁₆" arbor shank** McMaster-Carr #4008A581
» **Pilot drill bit, ¼", high speed** for the hole saw, McMaster-Carr #4066A89
» **WD-40** for use as cutting oil
» **High-speed rotary tool and miniature cutoff wheel, diamond grit, 1¼" diameter x 0.032" thick** McMaster-Carr #1257A89
» **Pipe tap, ¼-18 NPT, and handle**
» **Bolt, ¼-20, 1½"**
» **Blowtorch, MAPP gas**
» **Silver solder and silver solder flux** with the highest melting point you can find at your local hardware store
» **Helping hands tool**
» **Screwdrivers**
» **Wrench set, open end**
» **Wire strippers**
» **Wire crimps**
» **Wood saw**
» **Thread-locking compound, high strength**
» **Sandpaper, fine grit**
» **Metal file**
» **Hobby knife**
» **Two-part epoxy, 24 hour** It should outgas less than faster-curing epoxies during fusor operation.
» **Center punch**
» **Vise**
» **Masking tape**
» **Teflon tape**
» **PVC pipe glue**
» **PVC pipe, 1" nominal, about a 12" length**
» **Gloves, latex (2 pairs)**
» **Safety goggles**

For Those About to Fuse, We Salute You

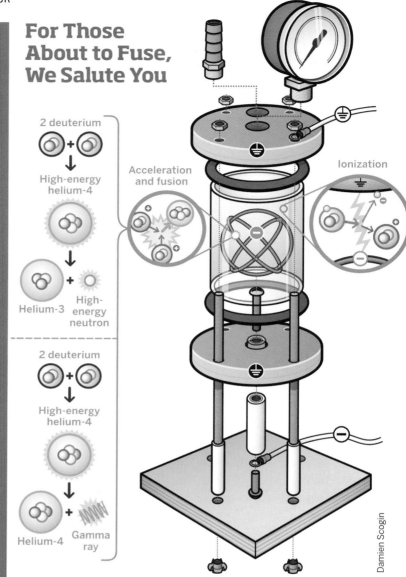

2 deuterium

↓

High-energy helium-4

Helium-3 + High-energy neutron

Acceleration and fusion

Ionization

---

2 deuterium

↓

High-energy helium-4

↓

Helium-4 + Gamma ray

Damien Scogin

How it works

The typical Farnsworth-Hirsch fusor has two concentric electrical grids inside a vacuum chamber: an inner grid charged to a high negative potential, and an outer grid held at ground potential. Our benchtop version has a stainless steel wire inner grid, and uses the aluminum chamber walls as the outer grid.

A variac controls the AC mains voltage input to a neon sign transformer, which steps up standard 110V AC to the 10kV range. A homemade rectifier converts AC to DC power to charge the grid.

A vacuum pump evacuates the chamber to a pressure of about 0.025mm of mercury, clearing the playing field so the few remaining gas molecules can accelerate without premature low-energy collisions. A vacuum gauge indicates the pressure inside.

High voltage across the grids causes gas molecules to ionize; that is, they lose an electron and become positively charged. Electrostatic forces then accelerate the ions — mainly O_2^+, N_2^+, Ar^+, and H_2O^+ — toward the high negative charge at the center. Some ions collide; those that miss the first time are arrested by the electric field and re-accelerated toward the center for another go.

Low-power fusors produce a beautiful purple ion plasma "glow discharge" similar to plasma globes and neon signs. In high-power fusors, the inertia of the ion collisions squeezes hydrogen atoms tight enough to fuse, hence the term *inertial confinement*.

High-power fusors typically fuse deuterium (D or ^2H) into helium and tritium. Deuterium is a hydrogen isotope whose nucleus contains a neutron in addition to the usual single proton. It occurs naturally in very low concentrations, primarily as hydrogen deuteride (HD) but also as "heavy water" (D_2O), "semiheavy water" (HDO), and deuterium gas (D_2). Only 1 in 6,000 hydrogen atoms is deuterium. Tritium (a hydrogen atom with *two* neutrons and one proton) is even rarer.

When two deuterium atoms fuse they create a high-energy helium-4 atom, which stabilizes itself by releasing a proton, a neutron, or a gamma ray. This release leaves behind a tritium atom, helium-3 atom, or helium-4 atom, respectively.

Fusor Nation

The fusor was developed in the 1960s by Philo T. Farnsworth, who also invented television. It's popular with DIY experimenters because it's easy to build and can reliably produce fusion reactions.

Fusors have yet to produce useful power, but they can be dangerous. They require high voltages and can produce harmful ultraviolet, x-ray, gamma, and free neutron radiation.

IEC reactors are currently being studied at MIT, the University of Wisconsin-Madison, University of Illinois, Los Alamos National Laboratory, and EMC Corporation, among other labs.

1a

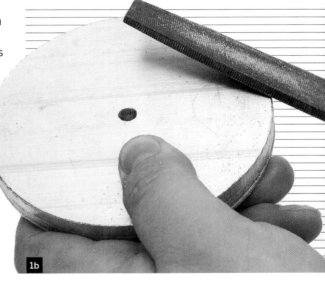

1b

Gunther Kirsch

Let's Do This
1. Cut the chamber parts.

Use the hole saw to cut two 4"-diameter blanks from the aluminum bar stock (**Figure 1a**). Use the slowest speed on your drill press and use WD-40 as cutting fluid. It's still going to make a lot of horrible noise. Clean up rough edges with a file, taking care not to scratch the surface of the blanks (**Figure 1b**).

Print and cut out the flange templates, provided online at makezine.com/36. Stack the 2 aluminum blanks with the template on top and bolt them together concentrically with a 1½"-long ¼-20 bolt. Use masking tape on the contacting surface to prevent scratches, and to hold the template in place

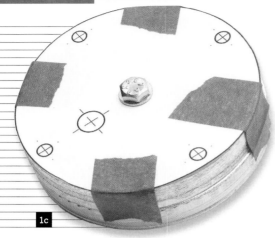

1c

1d

(**Figure 1c**). Center-punch all holes (**Figure 1d**) and remove the template.

Drill out the 4 bolt holes (leave the offset hole for later), first with a ⅛" bit and then again

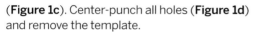

1e

1f

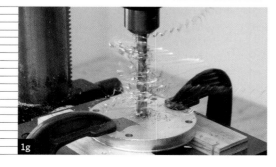

1g

with a ¼" bit (**Figure 1e**). This is called *step drilling* and it makes drilling large, accurate holes much easier. Make sure the blanks don't shift when drilling. Before you separate the flanges, use a center punch or marker to make witness marks on their edges so you can line up the 2 hole patterns correctly every time you assemble the fusor chamber (**Figure 1f**).

Drill out the center bottom flange hole to ⅜" (**Figure 1g**) and test-fit the ceramic tubing. It should be a close fit. If it's too tight, tape a strip of sandpaper to a nail, wrap the sandpaper tightly around the nail, chuck it in a power drill, and use it to expand the hole just enough that the ceramic tube fits snugly.

On the top flange, drill the center hole and the offset hole to 7⁄16" using the step drill method (**Figure 1h**). Clamp the flange in a vise, using tape or shims to prevent marring, and carefully tap the two 7⁄16" holes you just drilled to ¼-18 NPT (**Figure 1i**). Note that this is a tapered thread.

Following the templates provided online, use a hobby knife to cut 2 gasket rings (2½" ID × 3⅛" OD) from sheet rubber. (**Figure 1j**).

Use a rotary tool and diamond cutoff wheel to cut a 2" length of ceramic tubing (**Figure 1k**). Sand the rough edges smooth (**Figure 1l**).

2. Install the feed-through.

Using 24-hour epoxy, glue the aluminum standoff into the ceramic tubing, taking care not to get any glue in the threads (**Figure 2a**). You may need to file or sand down the outside

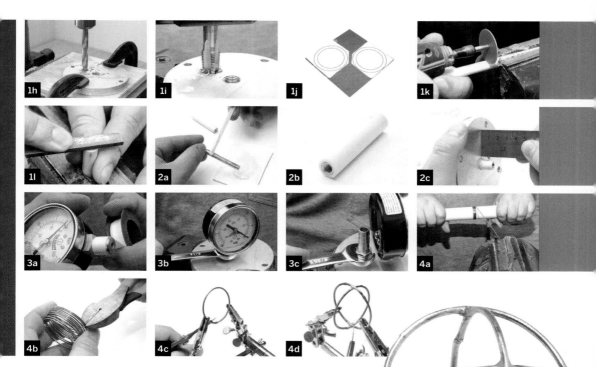

of the standoff to make it fit. Let the epoxy dry 24 hours. This is your high voltage feed-through (**Figure 2b**).

Using a healthy amount of epoxy, glue the feed-through into the center hole in the bottom flange, making sure it protrudes ½" into the chamber (**Figure 2c**). Again, keep glue out of the threads.

3. Add the vacuum ports.
Apply Teflon tape to the threads (**Figure 3a**) and then thread the vacuum gauge into the center hole in the top flange (**Figure 3b**) and the barbed hose fitting onto the offset hole (**Figure 3c**). Tighten securely with a wrench.

4. Fabricate the grid.
Cut a 12" length of 1" PVC pipe and then drill a ¹⁄₁₆" hole in the middle. Cut 48" of stainless steel wire and anchor one end securely in a vise or clamp. Thread the other end into the hole in the PVC pipe and neatly roll the wire around the pipe while applying constant firm tension (**Figure 4a**). Release the wire from the anchor point and clip it off the pipe, being careful to keep the wire from springing back. You'll be left with a small coil of wire with 10–12 turns, about 1⅝" in diameter (**Figure 4b**).

Clip 3 rings off the coil. Solder one ring closed using silver solder and a MAPP gas blowtorch (**Figure 4c**). Link a second ring through the first, solder it closed, and carefully solder the 2 rings together so they're at right angles (**Figure 4d**). Stretch the third ring over the first 2 to form a spherical cage with 8 equal openings. Trim it to size and solder it in place.

Finally, solder a #10-32 stainless steel machine screw to the outer ring, midway between two of the existing solder joints (**Figure 4e**).

5. Make the base.
Cut an 8"×8" square of ½" plywood. Use one of the flanges as a guide to lay out the 4 bolt holes. Drill ¼" holes and hammer in the 4 T-nuts. Adhere a rubber foot in each corner, on the same side as the T-nuts. (**Figure 5, following page**).

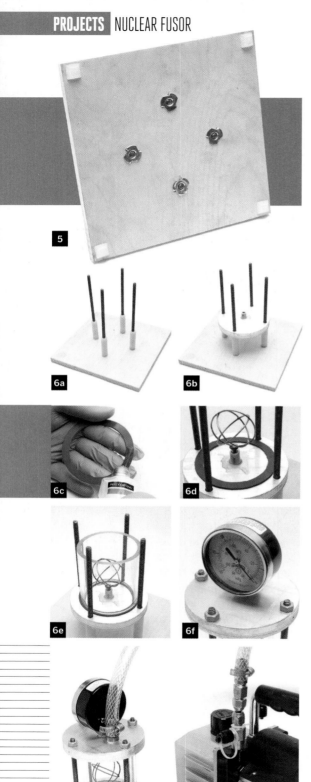

5

6a

6b

6c

6d

6e

6f

7

6. **Assemble the fusor chamber.**

Apply thread-lock to the threaded steel rods and screw them into the T-nuts until they're flush with the T-nut flanges, as seen in **Figure 5**. Slip a nylon spacer over each rod (**Figure 6a**), then stack the bottom flange on top of the spacers; make sure the longer end of the high-voltage feedthrough protrudes downward (**Figure 6b**).

Wearing latex gloves, wipe the surface of the flange with alcohol and wait for it to dry. Apply vacuum grease to both sides of a rubber gasket (**Figure 6c**), carefully center the gasket on the bottom flange, and thread the spherical inner grid cage onto the end of the high-voltage feedthrough (**Figure 6d**).

Wearing a fresh pair of gloves, clean the glass cylinder with alcohol and carefully place it on top of the gasket (**Figure 6e**). Apply vacuum grease to both sides of the second gasket and lay it on top of the glass cylinder. Try to avoid getting any grease on the sides of the glass.

Now clean the top flange with alcohol and stack it on top of the second gasket, making sure the flange witness marks line up. Finish it off with four ¼-20 nuts (**Figure 6f**). The nuts should be just finger-tight, only slightly deforming the rubber gasket. Overtightening them may crack the glass!

7. **Plumb your fusor.**

Cut a 2' length of ⅜" ID reinforced vinyl tubing, slide the 2 hose clamps onto it, and fit the hose over the barbs on your vacuum pump and fusor chamber. Tighten the clamps over the hose and barbs (**Figure 7**).

Follow the instructions provided by the manufacturer of your vacuum pump to get it operating correctly (vacuum oil, power, venting, etc.). Vacuum pumps vary, and you may have to install additional plumbing to fit a male barbed adapter sized for ⅜" ID tubing.

8. **Make the high-voltage rectifier.**

Solder the 2 high-voltage diodes together, making sure the ends with the white bands are facing away from each other (**Figure 8a**). Cut three 2" lengths of 16 AWG stranded wire, strip

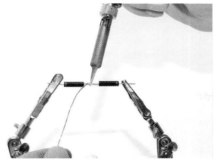

8a

8b

¼" of insulation off the ends, and bend each wire into an S shape. Solder one wire to each free end of the diodes and the third wire to the joint between the two (**Figure 8b**). Crimp a ring terminal to the free end of each of the 3 wires (**Figure 8c**).

Drill a #21 (or 4mm or ⁵⁄₃₂") hole into the top center of three ½" PVC pipe caps and "tap" the holes with a #10-32 screw (**Figure 8d**).

Feed the diode assembly through a ½" PVC tee union so that the middle wire goes out the side junction and the other 2 wires go out the ends (**Figure 8e**). Cut three 1½" lengths of ½" PVC pipe and glue them into the ends of the tee, around the wire leads (**Figure 8f**). Pass a #10-32 machine screw through each ring terminal and screw it into the threaded hole in each cap (**Figure 8g**). Glue the 2 side caps on, but not the top one yet (**Figure 8h**).

Place the tee in a vise with the opening facing up, and carefully pour in mineral oil until it's full to the bottom of the branch (**Figure 8i**). Loosen the vise, rock the tee gently side to side to make sure there aren't any trapped air bubbles, and then fill it all the way up. Screw the remaining ring terminal to the inside of the top cap and glue the cap in place, sealing the tee permanently (**Figure 8j**).

9. Wire the transformer.

If you're lucky, your neon sign transformer has a power cord attached. If not, open up a male plug with a ground prong and wire one end of a 3-wire AC power cord into it. The green wire goes to the ground prong. The black and white wires go to the hot and neutral prongs — it doesn't matter which is which (**Figure 9**).

On your NST there should be 2 large ceramic standoffs; these are your high-voltage outputs. Ignore them for now. We're interested in the 2 smaller input terminals and the single

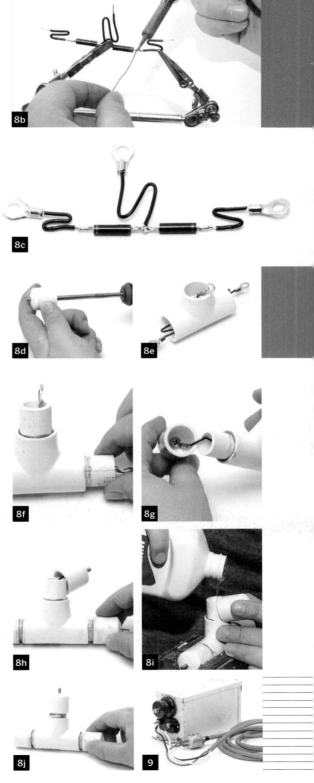

8c

8d

8e

8f

8g

8h

8i

8j

9

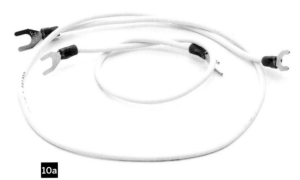

10a

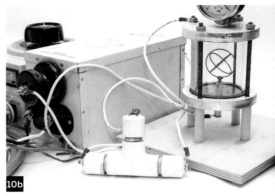

10b

ground screw.

Separate the 3 cord wires so the black and white wires reach the input terminals and the green wire reaches the ground screw. Strip ¼" of insulation off the wire ends and crimp a spade terminal to each, then affix the wires to the corresponding terminals. Again, the orientation of the black and white wires doesn't matter — it will work either way.

10. Wire the fusor.

Cut three 12" lengths and one 24" length of high-voltage wire, strip ¼" of insulation off the ends, and crimp a spade connector onto both sides of each one (**Figure 10a**).

Attach a 12" wire between each of the 2 high-voltage terminals on the NST and one of the 2 side terminals on the rectifier. Connect the third 12" wire between the top of the rectifier and the fusor's high-voltage feed-through using a ½"-long #10-32 screw.

Finally, connect the 24" wire to one of the top studs holding the fusor together, using the nut to hold it on the threaded rod. Attach the other end to the ground screw on your NST (**Figure 10b**).

11. Test the vacuum.

When working with glass vacuum chambers, always test the chamber from behind a safety barrier first. I used a door with a window and just ran the power cord under the door.

Plug in the vacuum pump and watch from safety as the needle on the vacuum gauge goes to 0. Leave the pump running for 5 minutes. If it doesn't implode during that time, it should be OK for normal use if treated gently.

Turn off the vacuum pump and allow the system to return to ambient pressure before handling or storing the fusor.

Running Your Fusor

Plug the transformer's power cord into the outlet on your variac, then plug the variac into the wall outlet. Don't switch the variac on yet.

Turn on the vacuum pump, wait for the gauge to reach 0, then wait another 2–5 minutes to reach deeper vacuum. Leave the pump running. Now switch on the power for the variac and slowly turn the knob, increasing the voltage fed through the transformer.

If everything works, you should see a bright purple discharge inside the chamber, and as you turn up the voltage a defined plasma ball will form inside the grid, with the occasional plasma beam leaching out through one of the grid openings. If you've built carefully, you may achieve the coveted "star in a jar": a glowing plasma ball with fine plasma lines radiating out in all directions through the grid openings.

Congratulations, you have successfully built a demonstration fusion reactor based on the principle of inertial electrostatic confinement!

WARNINGS:

Glass vacuum chambers can implode. Do not operate the vacuum system without safety goggles. Reminder: High voltage and current can injure or kill. Fusors may generate harmful radiation. Do not attempt to build or operate this fusor unless you understand the risks and are capable of safely using high voltage and vacuum equipment. Do not run this device at more than 12kV rectifier input.

Working With High Voltage

Our fusor is a relatively low-power version, with its components grounded to minimize the danger of shock. Still, accidents happen. Here's what you need to know to stay safe.

Electricians have a saying: Volts hurt, amps kill. *Current* is more dangerous than voltage — just 10–20 milliamps (mA) of alternating current (AC) can cause muscle contractions that prevent you letting go of the electrified object, and 70mA–100mA can cause heart fibrillation and death.

Household AC wall power is typically 120 volts at a lethal 15 amps of current. And it alternates at 60Hz frequency, which can also cause fibrillation. It's very hazardous.

The variable transformer modulates wall power up to 140V AC at 5A current — lower, but still a real electrocution hazard.

Next, the neon sign transformer steps up the voltage to a high 12,000V AC and slashes the current to 30mA — unlikely to cause fibrillation but still above the "let-go threshold."

Finally, the rectifier converts AC to DC — direct current — at 6,000V, 30mA. That's half the voltage of the NST output, and falls below the let-go threshold for DC current (about 75mA). But it can still be deadly, as DC causes worse contractions and tissue burns than AC.

So be sure to wire your fusor correctly, and avoid touching any part of it during operation except the variable transformer knob. And when in doubt, ask an expert before proceeding.

—Keith Hammond

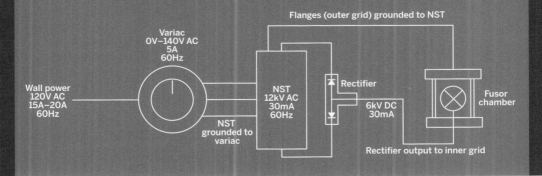

Never run your fusor for extended periods of time. A minute or two is plenty. Plasma beams escaping from the core may spot-heat the glass and cause it to implode. Make sure that you and everyone nearby wears ANSI-approved safety goggles whenever the cylinder is under vacuum.

Troubleshooting

When you first ignite your reactor, you may see sparks and arcs on the inner grid. As long as these don't persist in one spot, it's fine — just bits of dust and debris burning off, and after a while the sparks will stop. But if arcing persists, you've probably got carbon deposits, which

will continue to arc. This can be dangerous, so stop and clean the chamber thoroughly before proceeding.

If the glow discharge is deep purple, the seals are probably leaky; check the gaskets and reapply vacuum grease. With a good vacuum, you should see a bright, almost-blue purple glow. ◪

＋ Download templates for cutting and drilling the chamber flanges and gaskets at makezine.com/36.

＋ Forums, tips, and all things fusor: fusor.net

Dan Spangler is the fabricator for MAKE Labs.

Desktop
Foundry

Cast metal right on your desktop!

Written and
photographed by
Bob Knetzger

⚡ **TIME: 4–8 HOURS**   ⚡ **COST: $40–60**

MATERIALS

- » **Wood board, 7½"×5½"** I used a piece of ¾" solid oak shelving.
- » **Wood block, about 2"×2"×2½"**
- » **Soft brass wire, 16 gauge, about 2'**
- » **Brass tubing, ³⁄₁₆", 5" length**
- » **Brass strips, 0.030"×¼"×1½" (2)**
- » **Setscrew collars, ³⁄₁₆" ID (2)**
- » **Brass brads, round head, small**
- » **Glass bottles or vials, small** about 5ml
- » **Thimble**
- » **Alcohol lamp, about 1oz**
- » **Hardwood dowels: 1" diameter, 6" length (1); ¼" diameter, 3" length (1)**
- » **Wood screw and washer**
- » **Corks (2)**
- » **Sugru molding compound** item #MKSU1 from the Maker Shed, makershed.com
- » **Field's metal**
- » **Self-adhesive rubber feet**
- » **Kitchen matches and striker strip**
- » **White glue**
- » **Wood stain**
- » **Nonstick cooking spray**

TOOLS

- » **Hand drill or drill press**
- » **Drill bits: 1½", 1", ⁷⁄₃₂", ¼", ³⁄₃₂", ¹⁄₈", ¹⁄₁₆"**
- » **Wood saw**
- » **Screwdriver**
- » **Ruler**
- » **Locking pliers** aka vise-grips

OPTIONAL:
- » **Router with ogee curve bit**
- » **Mill and lathe**
- » **End mills: ¾" and ½"**
- » **Micro-Mark Pro-Etch etching kit** item #83123, micromark.com, if you're making the decorative brass parts. Contains 0.005" brass sheet, laminator, etching and stripping chemicals, etching tank and agitator, inkjet printer film, photo resist films, developing tray, measuring cup, safety goggles, and gloves.

Here's another "meta-project:" a DIY project that in turn makes other projects. It's a miniature, working foundry that casts real metal parts safely on your desktop. Make custom jewelry, tiny trinkets, diecast-style game tokens – then remelt them and recast, again and again.

What makes it all possible is a special eutectic alloy, Field's metal, which melts at an amazingly low 144°F (about the temperature of hot coffee). Unlike other low melting temperature metals, this alloy of bismuth, indium, and tin contains no lead or cadmium and is safe and nontoxic.

The basic foundry is made from wood and metal along with a few scrounged household parts. If you're up for a challenge, you can also dress up your Desktop Foundry with some snazzy brass trim and a twirling phoenix turbine, just for fun.

The simple foundry design has a center shaft mounted vertically on a wooden base. The shaft swivels 90° to move the crucible from the heating lamp over to the mold. The cork handle swivels to tilt the crucible and pour the molten metal into the mold below. There's also storage for molds, metal, and matches. This tiny foundry uses a thimble for a wick snuffer.

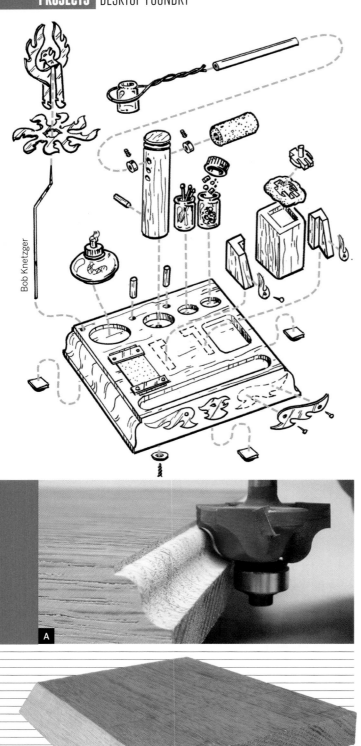

Bob Knetzger

1. Get the stuff!

None of the materials or dimensions for this project is critical except for getting your hands on some Field's metal. I found some for sale online through a scientific toy company. In small quantities it's a little pricey, but you'll only need a thimble-full or two of the stuff to have fun casting, melting, and recasting. Fun fact: It's named for its inventor, MAKE contributor Simon Quellen Field!

Use Sugru, the super easy-to-use silicone rubber material, to make the molds. It comes in handy small pouches perfect for this application.

I pilfered an alcohol lamp from an old chemistry set. For the see-through crucible and storage vials, I used contact lens bottles but any small glass bottles will do. To make it easy to pour, use a bottle with the least "shouldered" neck.

2. Build the base.

Cut the base to size. If you like, use a router to add a decorative flourish to the edges. I used an ogee curve bit for a "desktop pen set" look (**Figure A**). I also cut a 45°-beveled front face (**Figure B**), but you can leave yours plain.

Mark and drill the holes for the shaft (1"), your bottles, and your lamp (**Figure C**). All holes are drilled ⅜" deep, so use a drill press for best results. Drill a ⅛" through-hole in the center of the recessed shaft hole. I also milled some recessed tray areas for storage of the snuffer and extra molds, although you could use a router as well (**Figure D**).

Make a mold holder about 2½" high using 2×2 or any leftover

A

B

C

D

E

F

G

H

I

wood scraps from your shop. I milled a small recess in the top to help hold the Sugru, but it's not required (**Figure E**). Cut some L-shaped supports as shown here (**Figure F**), then cut bevels on the front. When glued in place in the center of the base, they'll keep the mold holder in position while casting.

Cut the 1" dowel to 5" long and drill a ⅛" pilot hole in the bottom for the wood screw. Drill a series of 7/32" holes crosswise through the upper portion of the shaft for adjusting the height of the crucible. Twist the dowel 90° and

drill a ¼" hole ⅜" deep, located ½" from the bottom of the shaft. I added some decorative rings using a lathe (**Figure G**).

Cut three 1" lengths of ¼" dowel and glue one into the hole near the bottom of the shaft. Place the shaft in the center hole in the base and fasten the wood screw and washer through the bottom (**Figure H**). Tighten the screw just snug, so the vertical shaft can rotate in the hole. Add 3 stick-on rubber feet to the bottom of the base (**Figure I**).

J

K

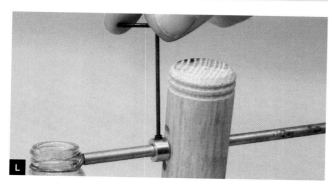

L

M

N

3. **Craft the crucible.**

Cut a 14" length of brass wire and wrap the center around the lip of the small glass bottle. Twist the wire to make a long handle, then cinch it tight on the bottle using vise-grips (**Figure J**).

Thread the twisted wire through the brass tube (**Figure K**). Put a setscrew collar over the tube, then slide the tube through one of the 7/32" holes in the shaft. With the alcohol lamp in place, turn the shaft and slide the brass tube so that the crucible bottle is directly over the lamp's wick. Hold the collar against the shaft and tighten the setscrew on the collar (**Figure L**).

With the crucible positioned over the lamp, hold one of the ¼" dowel pieces vertically on the base so that it touches the ¼" dowel in the shaft. Mark that position on the base and drill a ¼" hole ⅜" deep. Glue the second ¼" dowel in the hole (**Figure M**). This makes a positive stop to locate the crucible over the lamp.

Add the second setscrew collar on the tube and tighten it in place against the other side of the shaft. Drill a ¼" hole in the center of a cork and twist the cork snugly onto the end of the brass tube to make an insulated handle (**Figure N**). The crucible should swivel as you turn the cork.

4. **Mount the mold holder.**

Next, swivel the shaft 90° counterclockwise and position the mold holder underneath so that the crucible will pour directly into it. Carefully place the mold supports on either side and mark their positions on the base (**Figure O**).

Also hold a small dowel next to the dowel on the shaft, mark its position and drill a ¼" hole ⅜" deep. Glue the third ¼" dowel in place — that will make a positive stop for the shaft in the pouring position (**Figure P**). Glue the mold supports to the base, too.

5. Finishing touches.

Cut a 10" length of brass wire and wrap the center around the lip of the thimble. Using vise-grips, twist the brass wire tightly to make a handle as before (**Figure Q**). Drill a ³⁄₁₆" hole through a cork and thread it over the end of the twisted wires to make an insulated handle.

Finish the wood parts with a thin coat of dark wood stain to bring out the grain.

Drill some ¹⁄₁₆" holes in each end of the 2 small brass strips. Cut a piece of striker material from the side of a box of kitchen matches. Place the striker on the base as shown and put the brass strips on each end (**Figure R**). Mark the holes, drill ¹⁄₁₆" pilot holes and then gently tap in the small brass brads, holding the striker material to the base with the brass strips.

6. Make a mold.

Find a small coin, trinket, or object you'd like to mold. I sculpted a tiny MAKE robot from a piece of styrene. Spray some nonstick cooking spray onto the top of the mold holder (you don't want the Sugru to stick to it). Open up a pack of Sugru and knead it, then form it into the mold base. Spray your object with nonstick spray as a mold

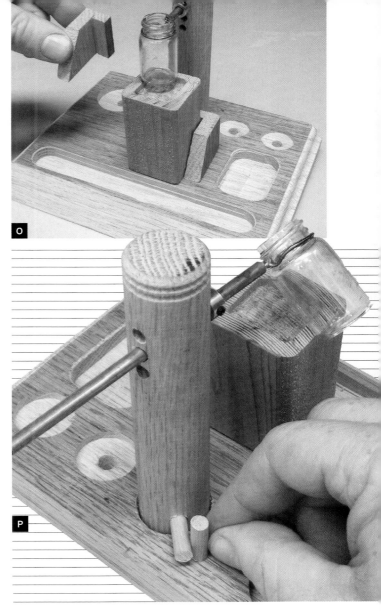

O

P

Q

R

S

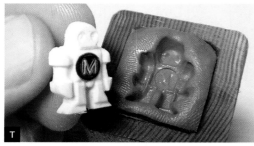

T

U

V

release, wipe away the excess, then carefully press the object into the Sugru (**Figure S**). Leave it in place for about 24 hours for the Sugru to firm up.

Carefully remove the object; you'll have a mold with every tiny detail and surface texture reproduced in it (**Figure T**). The Sugru will stay flexible and forgiving so you can easily unmold your cast metal shape, even with a bit of an undercut!

7. Cast a part!

Clean the oil from the mold, and dust it with a tiny bit of talcum powder to help the molten metal flow better. Put your mold in place on the base.

Place a small quantity of Field's metal in the crucible and swing it into position. Light the lamp. Gently twist the cork handle to rock the crucible back and forth as the metal melts. When ready, swivel the crucible over to the mold and twist the handle to pour the metal into the mold (**Figure U**). Give the mold some gentle taps to help the molten metal flow into details and to release any bubbles.

When cooled, flex to unmold the metal part (**Figure V**). You've cast a real metal treasure! You can remelt and recast again and again.

8. Get fancy (optional).

If you like, add some brass trim (**Figure W**) and a whirling phoenix turbine to make your Desktop Foundry worthy of your office or den! You can download my templates at makezine.com/36.

If you're lucky enough to have a local shop with a "fiber" laser cutter you could have some brass parts laser-cut for you. But many shops will not cut brass — it's too reflective and can damage their laser's lens. Instead I made detailed brass parts by chemical etching. It's similar to etching a PC board except there's no "board."

Photo-etching brass takes several steps: Print out your artwork on clear film and place it over the photosensitized brass (**Figure X**). Expose it to a light source, then remove the art and place the brass in developer. The areas covered by the black image wash away,

W

X

leaving an acid-resistant pattern. Place it in ferric chloride solution to etch away the unprotected areas, leaving your brass part (**Figure Y**). I used Micro-Mark's terrific all-in-one etching kit, which comes with complete instructions. (Watch for a future article on that!)

Trim and assemble the turbine as shown, and mount it to the base atop a pointed wire (**Figure Z**). The rising air from the burning lamp makes the flaming phoenix spin around.

Attach the etched brass trim parts to the base using small brass brads. I painted the areas underneath first with flat black paint for best contrast with the shiny brass. ◪

Y

See a video of the Desktop Foundry in action at makezine.com/36. You can also download files with artwork patterns for making the etched brass parts.

Bob Knetzger (neotoybob@yahoo.com) is an inventor/designer with 30 years of experience making fun stuff.

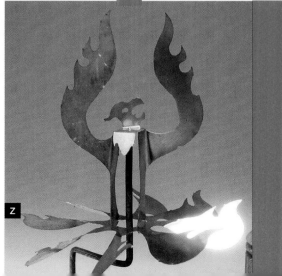

Z

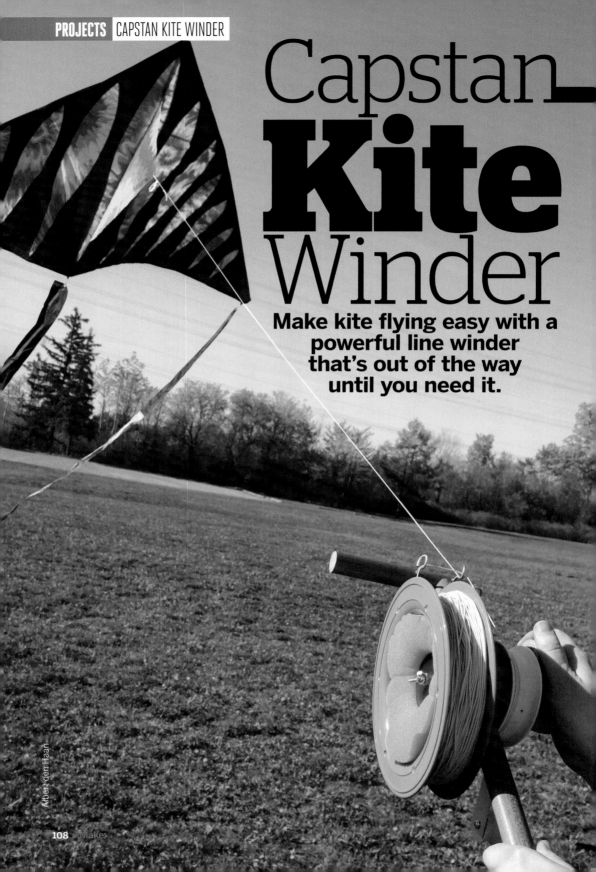

Capstan
Kite
Winder

Make kite flying easy with a powerful line winder that's out of the way until you need it.

Albert den Haan

Written by **Albert den Haan**

✎ TIME: A FEW HOURS   ✎ COST: $40–$50

➕ Handheld "halo" kite winders retrieve a lot of line with each turn, but they're hard to use with powerful kites. This DIY winder pulls in big kites fast using a capstan – a long crank on a small drum – to increase your leverage and protect the halo from the stresses of winding. It's easily built from hardware-store parts, breaks down for travel, and will even let you attach mechanical power.

Capstans require some tension on the loose "tail" of the line to prevent slippage on the drum. This winder uses a simple adjustable "slip clutch" to give the capstan the proper tension, so you can pay attention to the kite, not the winder.

Fly your kite with just the halo, then mount it on the winder for retrieval. The long double-T shape is easy to prop on a hip, and the crank gives you the choice of high-power or high-speed knobs. Use the winder to manage several kite lines at a time while doing your real flying with just the lightweight halo reels.

1. Build the handle.

Cut the dowel to make a main handle or "spine" the length of your arm from elbow to wrist, and 2 crosspieces 6" long.

Assemble as shown in **Figure A** (following page), using glue and #6 screws in ⁷⁄₆₄" predrilled holes. The top crosspiece should project about 1" on the capstan side; the bottom crosspiece is centered.

2. Make the capstan drum.

Saw the tire off the lawn mower wheel. Smooth any big bumps off the winding surface of the hub, but preserve any sub-millimeter texture to help the winding action.

If the axle bushing protrudes past the hub rim, cut it back flush. Press-fit the hose into the axle hole and trim it flush too.

Now cross-drill the axle bushing diametrically to accept a cotter pin. If your drill bit won't reach, clip a wire coat hanger to an angled point. You can use it to hand-drill soft plastic and wood.

Optional:
For extra strength, use a 1" drill bit to cut the spine. This creates a "bird's mouth" curve in each end for the crosspieces to seat into.

MATERIALS
- » **Hardwood dowel, 1" diameter or 1" square, 3'–4' long**
- » **Wood screws, #6×1½" (2)**
- » **Wood glue**
- » **Bolt, ⁵⁄₁₆"×4"**
- » **Screw eyes, ⅛"×¾"×2" (3)**
- » **Plywood or PVC foamboard, closed cell; ⅛"×8"×11"** such as Sintra or Lite-Ply
- » **Aluminum bar, ¹⁄₁₆"×1"×10"**
- » **Polyurethane foam, 1½"–2" thick** e.g. upholstery foam, enough to make a disc to fit inside your halo reel
- » **Lawn mower wheel, 6", with solid plastic hub** such as Arnold #490-320-0002. Look for a hub 4" or less in diameter, where the tire seats down inside the rim.
- » **Bolt, #6×1½", with nylock nut** or hex nut and spring lock washer
- » **Thin washers, ⁵⁄₁₆" ID (1–4)** various thicknesses, for shimming
- » **Hitch pin, ⁵⁄₁₆"×1-⁵⁄₁₆"** aka hairpin cotter pin or R-clip
- » **Wing nut, ⁵⁄₁₆"**
- » **Hex nut, ⁵⁄₁₆"**
- » **Flat washers, ⁵⁄₁₆" ID (4)**
- » **Flat washers, ³⁄₁₆" ID (2 or more)**
- » **Rubber washer, ⁵⁄₁₆" ID × 1½" OD**
- » **Plastic hose or tubing, ⁹⁄₁₆" OD × ¼" ID, 2" length** Most diameters smaller than ⁵⁄₁₆" should work.
- » **Shoelace or cord, at least 18"**
- » **Drawer pulls, round (2) with mounting screws** for crank handles
- » **Halo winder with kite line, at least 4" in diameter** IMPORTANT: The halo's smallest winding diameter must be larger than the mower wheel hub's diameter.
- » **Cotter pin, 1¾"**
- » **Graphite lubricant**

TOOLS
- » **Wood saw**
- » **Drill and drill bits: ⁵⁄₁₆" ⁷⁄₆₄" ³⁄₃₂" ⁵⁄₆₄"**
- » **Drill bit, 1" (optional)** or the same diameter as your dowel, for optional bird's-mouth joinery
- » **Screwdrivers**
- » **File**
- » **Hacksaw**
- » **Center punch**
- » **Vise**
- » **Scroll saw** for cutting discs
- » **Sandpaper**
- » **Utility knife**
- » **Pliers**

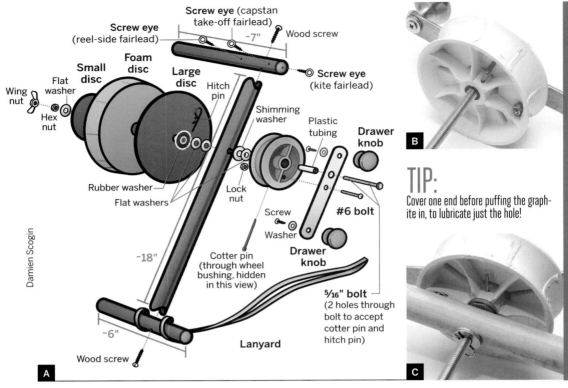

Screw eye (capstan take-off fairlead) ~7"

Screw eye (reel-side fairlead)

Wood screw

Foam disc

Small disc

Flat washer

Large disc

Hitch pin

Screw eye (kite fairlead)

Wing nut

Hex nut

Shimming washer

Plastic tubing

Drawer knob

B

Rubber washer

Flat washers

Lock nut

Screw

Washer

#6 bolt

Drawer knob

TIP:
Cover one end before puffing the graphite in, to lubricate just the hole!

~18"

Cotter pin (through wheel bushing, hidden in this view)

5⁄16" bolt (2 holes through bolt to accept cotter pin and hitch pin)

~6"

Lanyard

Wood screw

A

C

Damien Scogin

3. Make the crank.

Cut the aluminum bar to span your wheel hub and extend 3" past one side.

Drill holes for the knob screws ⅜" from each end. Locate and drill the 5⁄16" hole for the axle bolt so that the crank spans the hub completely.

Mount the knobs with 3⁄16" ID washers so they spin without wobbling or binding. The inner knob is for low-power cranking, the outer for high-power.

4. Prepare the axle.

Use a 5⁄16" drill bit to ream out the hose in the bushing so the 5⁄16" bolt can be inserted and removed with mild force.

Press the axle bolt through the crank handle and wheel hub, then use the hole in the bushing to mark the bolt for its cotter pin hole. Disassemble.

Center-punch your mark, mount the bolt in a vise, and cross-drill the bolt with a hole just big enough to pass the cotter pin, probably about 5⁄64".

5. Assemble the capstan.

Drill through the crank and hub, close to the rim, and install the #6 bolt with locking nut. This helps to carry all cranking torque to the wheel hub.

Install the cotter pin to lock the capstan wheel to the axle (**Figure B**).

6. Drill the handle.

Use the assembled capstan to correctly place the 5⁄16" axle hole in the upper part of the main handle, so that the crank (and your hand) are clear of the top crosspiece and its hardware.

Cross-drill a 5⁄16" hole for the axle bolt in the same plane as the top crosspiece. Lubricate the hole with powdered graphite.

7. Place the hitch pin.

Thread onto the capstan axle a flat washer, shim washer, main handle, and 2 more thick washers. Then mark the axle bolt for a hole to accept the hitch pin. Leave a bit of room between this hole and the last washer to account for the thickness of the pin.

Disassemble and cross-drill the bolt as you did for the cotter pin. Reassemble and insert the hitch pin in the new hole to hold the axle onto the handle. Add or swap shim washers as needed to take up slack (**Figure C**).

8. Make the halo adapter.

Cut 2 discs of ⅛" PVC or plywood: one just smaller than the hole through your smallest halo reel, another just larger than the inside of your largest reel. Sand the edges smooth and drill a ⁵⁄₁₆" center hole in each.

Cut a foam disc to fit very snugly inside your reel, and drill or punch a ⁵⁄₁₆" center hole.

On the axle, add the rubber washer, large disc, foam disc, small disc, flat washer, and ⁵⁄₁₆" hex nut. Tighten the nut so that all the discs rotate with the axle when cranking but can be stopped with the pressure of a bare thumb.

Now tighten the wing nut solidly against the hex nut. This helps prevent the slip clutch from loosening or tightening itself.

9. Add fairleads.

Twist the loops of the 3 screw eyes slightly out of alignment so that kite line can easily be slipped in diagonally but won't wander out. Twist at least one left and at least one right, so that threading the winder will be easier (**Figure D**).

Drill ³⁄₃₂" pilot holes and install the screw eyes as shown. Align the kite fairlead with the outermost flat surface of the capstan. Align the capstan takeoff fairlead with the innermost flat spot of the capstan; it should protrude upward slightly from the crosspiece. Align the reel-side fairlead with the centerline of your reel.

10. Add locking lanyard.

Tie the restraining lanyard to the bottom crosspiece so it can be looped over the high-power knob to lock the action.

Use It

To retrieve your kite, mount the halo on the foam adapter disc and thread the string backward, as follows:

- Hold the string on the top crosspiece with your capstan-side hand. Mount the reel on the foam disc so that the line reels off the bottom, and thread the line through the reel-side fairlead.
- Switch hands and thread the line through the capstan takeoff fairlead, around the capstan 3–4 times in the same direction as the halo reel, then out through the kite fairlead to the kite. You can see a video of this process at makezine.com/go/capstanthread.
- Adjust the fairleads as necessary by twisting. Now reel that big fella in.

For even more power, chuck a ½" socket in your cordless drill and use it to drive the axle bolt! ◪

Albert den Haan (hackingonkitebits.blogspot.ca) mixes sailing, kite flying, robotics, and everyday materials to fill Ottawa winter nights.

Albert den Haan

Shrink-Film Gaming Minis

Written by **Sean Michael Ragan**
Illustrations by **Nate Van Dyke**
Terrain built by **Tai Hake**

Gunther Kirsch

⚡ **TIME: 1–2 HOURS** ⚡ **COST: $10-$25**

MATERIALS

» **Shrink film, inkjet printable, 8.5"×11" (5)** I was disappointed with the "clear" type (which really isn't clear at all, after shrinking) so I recommend "white."

» **Binder clips, 1.25" width (12)** such as Acco "medium"

TOOLS

» **Kitchen oven**
» **Computer, inkjet printer, and ink**
» **Guillotine paper trimmer** or hobby knife
» **Scissors**
» **Cookie sheet**
» **Cardboard sheet** to fit cookie sheet
» **Paint pen (optional)** any color

Faster than figurines, tougher than cardboard, and way more fun to mak

✚ Even in an age of immersive video games, there are those who like to play tabletop games. Real-world wargaming is very different from playing a multiplayer online game, even with real-time voice chat. Whereas Eve Online or Xbox Live is kind of like hanging out with your buds watching TV, actually getting together and playing a board game is more like a real party. Manipulating the physical game pieces is also satisfying in a way that virtual objects have yet to achieve.

Many tabletop gamers eventually end up making their own pieces, for one reason or another. They may be making a custom army for an established game, or they may be inventing a game from scratch. The time-honored method is to use illustrated cardboard "chits" that lie flat on the table or stand upright in simple bases. For pieces fully "in the round," you can buy commercial 3D figures

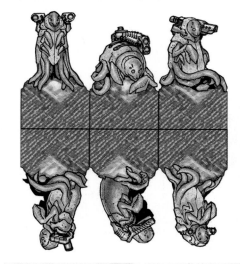

and customize them to your taste. If you have the time, you can sculpt your own minis and (these days) even 3D-print them.

Shrink film makes a nice intermediate approach to original mini design, midway between cheap cardboard cutouts and fully dimensional figurines. Shrink plastic is much more durable than cardboard and, unlike traditional minis, you can make your artistic mistakes at the software level, where they're easily fixed.

Feel like giving it a shot? Read on.

Designing Your Minis

Check the directions to figure out how much your film is going to shrink. The film we used shrinks by about 50%, meaning that the designs as printed need to be about twice as big, in each dimension, as the desired miniature size.

If you want to make your minis stand upright by adding bases, be sure to leave an empty "tab" at the bottom of each image so they can be attached without covering the art. Medium-size binder clips, for instance, have a footprint of about 1.25"×¾" which, accounting for 50% shrink, will cover a 2.5"×1.5" "tab" at the bottom of each image.

If you don't want to design your own minis, or just want to experiment with the method, we've put together two opposing sets of squad-level markers for Ground Zero Games' fantastic open-source sci-fi wargame *Stargrunt II*. Download the ready-to-print artwork from makezine.com/36.

1. PRINT 'EM OUT

Shrinking causes the color saturation of your images to increase, so they will need to be lightened before printing to keep them from ending up too dark. You can use the color settings dialog on some printers for this, or a dedicated image-editing program like GIMP or Photoshop. We've had best results from turning the brightness up about 50% before

1

2

printing, but you may want to experiment before printing a large batch.

2. CUT 'EM OUT

You can make the profiles as complex as you like, but keep in mind you may have to cut out a lot of them, so you may want to avoid a lot of intricate edge detail and internal openings. Square or rectangular is naturally easiest. Casualty and other "flat" counters can be designed this way, and cut out quickly in bulk using a swing-arm paper cutter.

3. SHRINK 'EM

Refer to the directions on your film for shrink times and temperatures. Ours took about 10 minutes per batch, in an oven preheated to 300°F.

Watch the pieces as they shrink. They will curl up, then re-flatten — sometimes not completely. If that happens, compress them gently with the back of a spatula while they're still hot. Let them cool before handling.

4. FINISH THE EDGES

I find a black edge looks better than a bare one. A paint pen works better than a Sharpie.

5. ADD BASES

Bases can be improvised from binder clips. The photo sequence illustrates the process. First, attach a clip to the bottom of each base, covering the empty tab at the bottom of the image. Then remove the wire handles from both sides of the clip by compressing them. The spring remains in place on the mini, forming the base.

Going Further

If your printer can align duplex printouts closely, it may be possible to print mirror images on either side of the film and make double-sided minis. You could put the same design on each side, for looks, or put an "injured" or "battle-damaged" unit on the back. ◾

Sean Michael Ragan is technical editor of MAKE magazine. His work has appeared in *ReadyMade*, *c't – Magazin für Computertechnik*, and *The Wall Street Journal*.

3

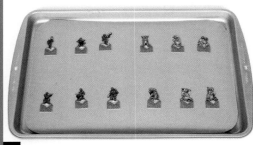

4

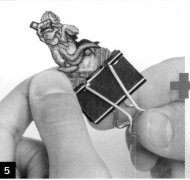

5

1.2.3. Smartphone Signal Generator

By *Jacob Beningo* Illustrations by *Julie West*

A SIGNAL GENERATOR IS HANDY TO HAVE AROUND THE LAB. It's perfect for testing inputs on a new hardware design; use it with an oscilloscope to verify the behavior of your circuit. Turn a smartphone into one for less than $15!

YOU WILL NEED:
» Stereo headphone plug, 3.5mm » Wire, 1-conductor, insulated, 28-gauge solid, 8" or custom length (3) » Insulated alligator clips, red (2) » Insulated alligator clip, black » Soldering iron with solder » Smart device with headphone jack

1. Wire the adapter.
Unscrew the cap from the 3.5mm plug. Strip ¼" of insulation off the ends of each wire. Solder one wire to the plug's large tab first — this will be the ground. Then solder each of the other wires to the smaller channel tabs — these will be the signal wires. Replace the cap. Either solder or screw the opposite end of the ground wire to the black alligator clip. Then connect the 2 red alligator clips to the signal wires. Make sure the alligator clip insulator is slid down the wire beforehand so they can be replaced after soldering.

2. Download signal generator app.
Download the appropriate signal generator app for your device. On iOS, Sig Gen is one option. On Android, Waveform Generator is a great app to use. These apps will generate various waveforms at different frequencies and amplitudes. They range in price from free to a couple dollars.

3. Connect adapter to smartphone.
Plug the adapter into the phone. Connect the clips to your oscilloscope, ground to ground and then signal to input. Turn headphone volume to maximum. Start the application. Select the waveform, amplitude, and frequency and start generating waves!

Going Further
To protect the headphone jack, a simple buffer can be added between the adapter and the circuit. For an example, and suggestions on how to use your signal generator, visit makezine.com/36. ⬈

Jacob Beningo (jacob@beningo.com) is a lecturer and consultant on embedded system design. He's an avid tweeter, a tip and trick guru, a homebrew connoisseur, and a fan of pineapple. What!

Laser Projection
Microscope

Laser pointer + drop of water = microbial movie theater!

Written by **Sean Michael Ragan**

⚡ TIME: 1 HOUR   ⚡ COST: $15 AND UP

➕ Inspired by Adam Munich's page about his DIY laser microscope, (makezine.com/go/teravolt), I built this project and discovered it's near the top of the list when it comes to getting the most bang for your buck. Whether you're a scientist, a science educator, or just a bright curious monkey, you really have to try this one for yourself. A common laser pointer, a drop of dirty water, and boom – giant microbes wriggling on the walls!

The tricky part is getting everything aligned – the laser, the projection surface, the hanging drop of water – but this simple stand built from junk-box odds and ends makes it easy. The laser and syringe are mounted to cheap hardware clips on super-magnets that allow for easy adjustments, but fix everything in position once you've got it tuned.

MATERIALS

- » **Heavy mounting base, about 6" long** Wood is suggested.
- » **Corner brace, galvanized steel, heavy duty 4"–6" legs with a triangular third side,** such as Stanley #755565
- » **Laser pointer** I used a green 30mW laser from Wicked Lasers; MAKE Labs had good results with cheap 5mW laser pointers. A continuous on/off switch is handy, but not necessary (and it's more hazardous to operate, because it won't turn off when you drop it). Common DPSS laser pointers work well.
- » **Hose clamps, vinyl coated (2)** to fit your laser pointer, such as KMC part #COV1109Z1
- » **Syringe with blunt needle, 5mL–10mL** Mine came from an inkjet cartridge refill kit.
- » **Broom clip, small** to fit your syringe
- » **Magnets, rare earth, disc or ring form (3)** I used Magcraft part #NSN0587.
- » **Washers, flat, #8 (1–4)** for spacing your laser from the bracket
- » **Liquid sample** I used stagnant water from a drainage ditch.
- » **Screws (2)** to mount bracket to base. MAKE Labs used pan-head wood screws with their wood base.
- » **Bottle cap**

TOOLS

- » **Pliers, needlenose**
- » **Scissors or drill**
- » **Screwdriver**
- » **Protective eyewear** appropriate to your laser

Gunther Kirsch

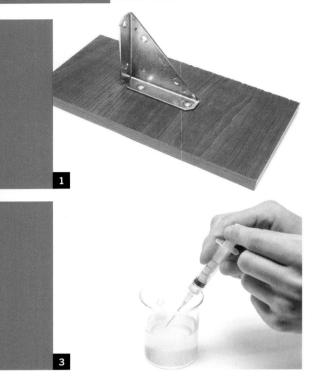

1. Mount the corner brace.

Cut a strip of mounting tape to fit one entire side of the corner brace. Adhere it to the brace and affix it to your base. Or drill the base and mount the brace with screws.

2. Assemble mounting clips.

The syringe clip consists of a magnet and a broom clip.

The 2 laser clips each consist of a magnet and a hose clamp. That's it!

3. Load the syringe.

I started out with scummy water from a drainage ditch, but eventually discovered that dirty water from a flowerpot was just as interesting.

Load the syringe with about 1mL of your sample by dipping the needle and pulling back on the plunger.

4. Assemble.

Attach the syringe clip to the brace's upright arm. Try not to let these magnets "snap" against the metal too hard; they're surprisingly easily to break.

CAUTION: Exercise extreme care when handling a syringe. Use a blunt needle with a cap, and leave the cap in place at all times when the syringe is not in use. If you don't have a cap, bury the tip of the needle in a wine cork.

Do not underestimate the potential dangers of stagnant and/or septic water. Use gloves when handling it, and wash your hands carefully when you're done.

TIP With some lasers, you can use the hose clamp to keep the laser's activation button depressed.

Mount the laser in its 2 clips and attach them to the triangular vertical plate. Then align the lens with one of the holes in the upright. Add washers between the magnet and the clamp if necessary to adjust spacing.

Mount the loaded syringe in its clip and align the tip of the needle just above the lens of the laser.

Set the bottle cap underneath the syringe to catch any drips.

5. Fire it up!

Put on wavelength- and power-appropriate protective eyewear for your laser. Position the projector on a flat, level surface, above waist height, about 5–10 feet from the projection surface. (The image is focused at infinity, so choosing a distance is a tradeoff between how big you want the projected image and how much power your laser puts out.)

Activate the laser, and adjust the position of the stand and/or laser clips to center the laser dot on the projection surface.

Depress the syringe plunger lightly to extrude a droplet of liquid at the tip of the needle.

Being careful to avoid looking directly into the beam, adjust the position of the syringe until the water droplet is perfectly centered in the beam path. This will be obvious from the projected image. When you've got it, the bright dot will mostly disappear and the microscopic image will appear. Enjoy the show!

GOING FURTHER

The earliest publication of this idea that I'm aware of is Gorazd Planinsic's 2001 article "Water-drop projector" in the journal Physics Teacher (free PDF at makezine.com/go/planinsic). On page 20, it provides a formula for calculating the approximate magnifying power of your projector based on the radius of the droplet, refractive index of the liquid, and distance to the screen.

It's been suggested that it may be possible to focus blue and red lasers on the same drop and achieve a 3D projection when viewed through red/blue 3D glasses. My experience with this monochrome projector suggests that getting such a system aligned correctly would be chal-

4

TIP If your image has a lot of noise that rotates with the laser housing, that probably indicates dust on the lens. Cleaning mine with a cotton swab moistened with rubbing alcohol did wonders for the image quality.

5

CAUTION: To be effective, this project requires a Class 3 laser in the 5mW–500mW output power range. These lasers present real hazards to vision, and should not be used without appropriate protective eyewear. Never look directly into the beam of a laser, and beware of the danger of reflection from shiny surfaces, such as the metal parts used in the projector itself.

lenging, but not impossible.

MAKE Labs tested this project with 5mW red, green, and violet pointers and found that the green performed best for brightness and projection distance. We couldn't tell if the different wavelengths affected image sharpness, but that might be more obvious with higher-powered lasers. Experiment and let us know what you find out at makezine.com/go/laserprojection! ↗

Sean Michael Ragan is technical editor of MAKE.

Try a Triac

Written by **Charles Platt**

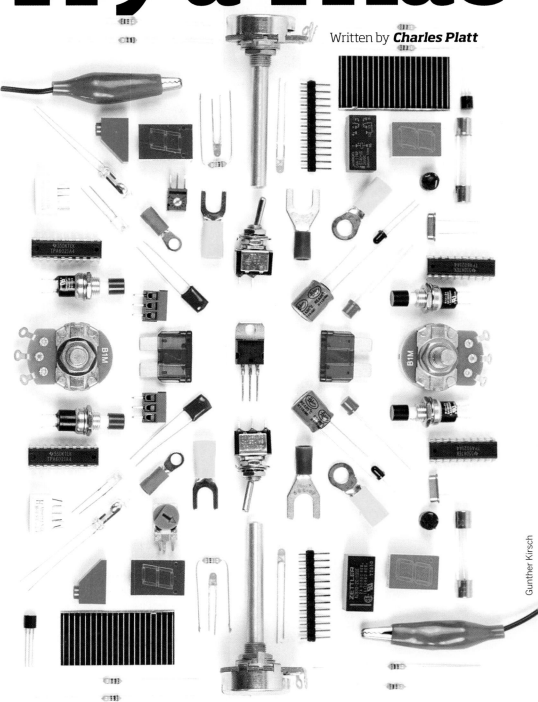

Gunther Kirsch

Clever off-label uses for an underappreciated component.

✎ TIME: 1 HOUR   ✎ COST: $10–$20

There are billions of triacs in the world. In almost every lamp dimmer, every electric stove, and many motor controllers, power is moderated by a triac clipping a portion of each positive and negative AC pulse.

When I started writing about this ubiquitous semiconductor for the second volume of my *Encyclopedia of Electronic Components*, I didn't expect to find much new to say. After all, the triac was invented more than 50 years ago. Imagine my surprise when I realized that it could have some low-voltage DC applications. Why, it could even be controlled by an Arduino! That was when I decided that a triac and myself should get personally acquainted.

Testing, Testing ...

By combining five segments of silicon, the triac achieves some unexpected capabilities. Like a transistor, it switches current. Unlike a transistor, it doesn't discriminate between positive and negative. You can run electrons through it in either direction, and it won't mind at all. Likewise, it will respond to either forward or negative bias on its gate terminal. It's also "regenerative," continuing to pass current even after the gate bias is removed.

Figure A shows a typical triac, while its schematic symbol is shown in **Figure B**. Sometimes the triangles in the symbol have open centers, and the symbol may be flipped or rotated. These variations make no difference. The gate is labeled G, while the main input/output terminals are labeled A1 and A2 (or sometimes T1 and T2, or MT1 and MT2). If the terminals are not identified in a schematic, A1 is always the one nearest to the gate, and gate voltage is always measured relative to A1.

So long as the gate has the same potential as A1, the triac blocks current in both directions. When the gate voltage swings higher *or lower* than A1, the triac will conduct current either way. Above a level known as the *latching current*, the flow will continue even if the gate voltage drops to zero. The flow continues till it falls below a level known as the *holding current*. These parameters are listed as IL and IH in triac datasheets.

A test circuit can be safely breadboarded, because although triacs are intended to work with 110VAC or

MATERIALS

» **Triac, 10mA gate trigger current** BTB04-600SL or similar
» **Trimmer potentiometers, 2kΩ (2)**
» **Resistor, ¼W, 330Ω**
» **Resistor, ¼W, 680Ω**
» **LEDs (2)** 20mA forward current
» **9V battery (2)** or split power supply, minimum +/– 9VDC

TOOLS

» **Breadboard** mini or full-size
» **Multimeter**
» **Hookup wire** or jumpers

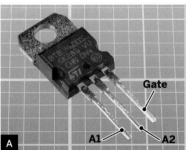

A typical triac, with gate and main terminals identified.

Charles Platt

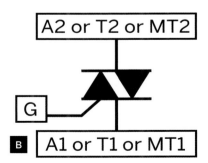

The triac schematic symbol resembles two diodes back-to-back, suggesting its functionality.

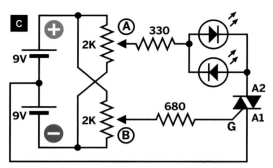

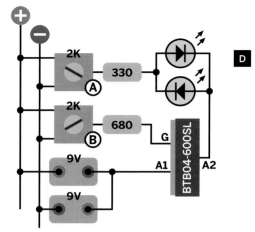

A simple low-voltage, DC test circuit uses two LEDs to show the current flow, and trimmers that apply current to the gate and between the main terminals.

A simple test layout for breadboarding.

higher, many will function with 12VDC or less, and can switch LEDs instead of light bulbs. I chose the BTB04-600SL because it will pass up to 4A of AC but can be triggered by as little as 10mA at 2VDC. Many triacs have similar specifications.

To provide positive and negative current, I used a pair of 9V batteries as in **Figure C**. You can substitute a split power supply if you have one. A 2K trimmer labeled "A" in the schematic applies a voltage ranging between +9V to −9V through a 330Ω resistor and a pair of LEDs to terminal A2 of the triac. Your LEDs should be rated for at least 20mA forward current. They are oriented with opposite polarities, to show which way current is flowing.

A second 2K trimmer, labeled "B," applies

+9V to −9V through a 680Ω resistor to the gate terminal. The resistor values were chosen to provide just enough gate current and latching current without burning out the LEDs. **Figure D** suggests a breadboard layout. **Figure E** shows the actual components.

Set both trimmers midway through their ranges, and connect the power. Turn trimmer "A" all the way toward the positive end of its range, and nothing happens yet, because trimmer "B" is applying neutral voltage to the gate. Now turn "B" either way, to apply positive or negative gate bias, and the triac will start passing current, lighting the top LED.

Things get interesting when you turn trimmer "B" back to its neutral position. This deprives the triac of gate voltage, but it still keeps

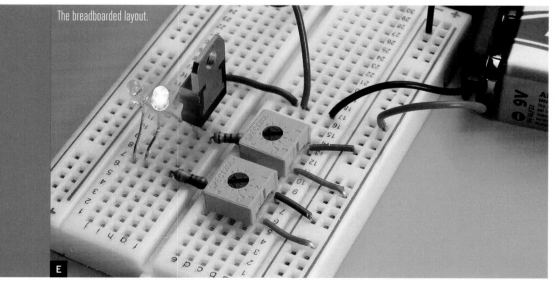

The breadboarded layout.

conducting, because the 20mA passing through it is just above its latching current. You can even unplug trimmer "B," and it makes no difference.

If the triac remains conductive with no gate voltage at all — how can we stop it? Simply turn trimmer "A" slowly back toward its neutral position, and when the current into terminal A2 drops below 10mA, the LED goes out.

If you repeat this test with trimmer "A" supplying −9VDC instead of +9VDC, the other LED will light up. It may not behave in exactly the same way as the first LED, because the triac's response is not completely symmetrical.

What Next?

The triac is a *latching* device, which makes it ideal for pushbutton activation. Send a pulse to the gate, and a motor powered through the triac will start running, and keep running. You can stop it by interrupting the power supply, or by bypassing the triac briefly to zero-out the potential between its main terminals. Add an appropriately wired DPDT switch, and your start button can make a 12VDC motor run forward or in reverse.

Try it yourself using the circuit in **Figure F**. It runs a motor, stops it automatically, and reverses it, all from a single button-press.

Pushbutton S2 energizes the triac, which starts the motor turning. At the end of its arc of rotation, a cam closes a limit switch (S3 or S4), which flips a latching relay (S1) to reverse the power. Most relays break one contact an instant before making the other. That hiatus without current will be sufficient to switch off the triac, which cuts power to the motor. The motor will stop a little way past the limit switch, so that the switch does not waste power continuing to energize the relay coil. When the triac is triggered again by S2, the motor now turns in the opposite direction. S5 will stop the triac at any point by bypassing it briefly, while S2 will restart it.

Some people like to trick out their cars with motors that open the hood or the trunk or do similar stunts. This circuit would be well

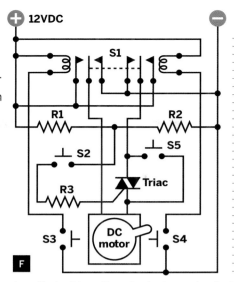

+ 12VDC −

F

A possible circuit to enable one-touch auto-reversing of a simple DC motor, for car accessories or home automation. S1 is a latching relay. S2 starts the motor. S3 and S4 are limit switches activated by the cam on the motor shaft. R1 and R2 could be 10K, functioning as a voltage divider to establish an initial gate voltage that is always different from the voltage on terminal A1. R3 would be chosen to provide appropriate voltage and current to the gate.

suited to that application. Similarly, linear actuators, such as those sold by Firgelli, can operate gadgets in the home like drapes that open or close or an entertainment center that emerges from inside a cabinet. Often these motors require 12VDC, which can be provided by cheap AC adapters designed for laptop computers. Here again a triac could be used for motor control.

Another possible application would be a "panic button" to stop a motor by interrupting power through a triac, in a battery-powered device such as a robot.

Lastly, because the gate of a triac only requires a few mA, it can be activated by a microcontroller. Check your triac's datasheet; it's probably Arduino-compatible.

For too many years, this quirky semiconductor has been ignored. What other applications could it have? Let your imagination be the limit switch. ◪

Charles Platt is the author of *Make: Electronics*, an introductory guide for all ages. He is completing a sequel, *Make: More Electronics*, and is also the author of Volume One of the *Encyclopedia of Electronic Components*. Volumes Two and Three are in preparation. makershed.com/platt

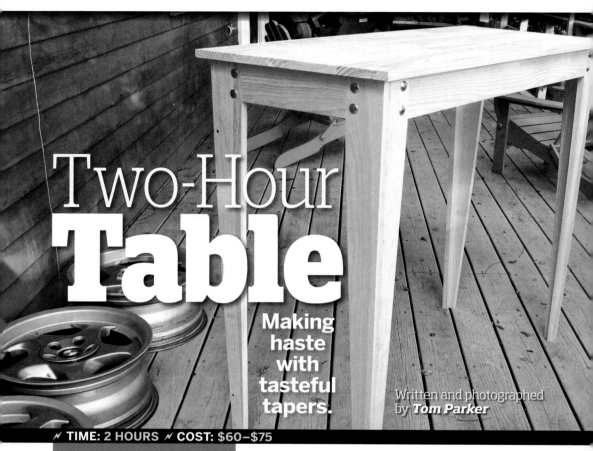

Two-Hour Table

Making haste with tasteful tapers.

Written and photographed by *Tom Parker*

⚡ TIME: 2 HOURS ⚡ COST: $60–$75

MATERIALS

- » **Solid pine panel, kiln-dried,** ¾"×20"×48" edge-glued and finger-jointed
- » **Pine boards, nominal 1×4, 8' long (7)**
- » **Carriage bolts, 2"×⅜" (16)** each with nut, flat washer, and lock washer
- » **Wood screws, 1¼" (20–24)**
- » **Wood glue**
- » **Brads, 16 gauge, 1½"**

TOOLS

- » **Chop saw** aka miter saw
- » **Table saw**
- » **Hammer, or brad nailer with compressor**
- » **C-clamps**
- » **Framing square** aka steel square
- » **Straightedge**
- » **Drill with twist bits and countersink bit**
- » **Screwdriver**
- » **Wrench or socket set**

➕ You've heard the old joke: A thing can be built fast, good, or cheap – pick any *two* of these qualities. Well, I like fast, good, *and* cheap. With this table I wanted all three.

I wanted an enduring shape with tapered legs that would age gracefully, hold plenty of weight, and weigh no more than a decent wooden chair. Using inexpensive, off-the-rack materials from a big-box lumberyard, I wanted to load up with supplies after work and have the table done before dark. That gave me about 2 hours from first flying sawdust to cleanup.

1. Shop.

I started with a shrink-wrapped, edge-glued and finger-jointed, ¾" thick, kiln-dried pine panel measuring 48"×20". For some projects these "convenience" panels are perfect. They're durable, they don't delaminate or

A

TIP: Time sands all wood, I say. Choose materials that can take a beating and age gracefully. That's why I avoid veneers and laminates. Solid lumber will withstand abuse (and refinishing) over the years. Oh, and I like wood-filler, primer, and paint for projects like this. Once painted, this table will last for centuries.

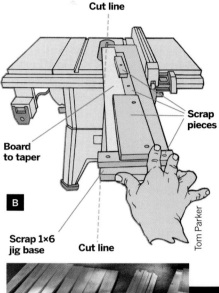

Cut line

Scrap pieces

Board to taper

B

Scrap 1×6 jig base   Cut line

Tom Parker

require edging like plywood, and they provide a square and accurate reference for building the rest of the table.

I also grabbed seven 8-foot pine 1×4s and sixteen 2"×⅜" carriage bolts with washers and nuts. You'll save big if you buy these bolts in bulk or find a farm supply store that sells loose hardware by weight, not by the piece.

2. Chop.

I rolled my table saw and chop saw out of the garage and onto the driveway to minimize indoor cleanup.

Clamping 4 of the 8-foot 1×4s together, I gang-cut all 8 pieces for the legs with just 3 chop cuts (**Figure A**). The length? Whatever you like, as long as all 8 match. I chose 41" so I'd have a few nice, square-ended scrap pieces to use for temporary mounting blocks later.

3. Make a taper jig.

I love taper jigs. They take only minutes to make and, for projects like this, you can eyeball the layout start to finish.

You'll run the entire jig through your table saw between the fence and the blade, with the table leg piece attached to it. The blade will remove the part of the leg extending beyond the edge of the jig. If the jig holds the leg at an angle, the result will be a tapered edge. The taper can be whatever angle you like, as long as the last 4" of the leg remains untouched by the blade. This keeps the leg tops square and parallel for final assembly with the tabletop.

To make the jig, I placed a leg piece on top of the jig's backing board at an angle that looked about right, being careful to keep the top 4" of the leg from extending beyond the edge of the backing board. Then I screwed down 2 scraps (of the same ¾" thickness) to backstop one side and the top end of the leg piece. Next, I added 2 extra scraps as hold-downs (**Figure B**).

4. Rip.

Once the jig was completed it took only a couple of minutes to run all 8 pieces through the table saw to taper them (**Figure C**).

C

D

TIP: Start with factory-squared materials when possible and make your project "self-referential" to minimize the need for measurements and custom cuts. A flat tabletop with true corners, off the shelf, makes its own project benchmark.

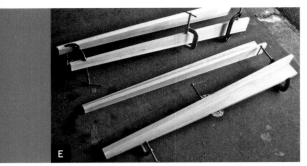

To make the finished legs symmetrical when the pairs are butted together, I chose 4 of the 8 pieces and ripped ¾" off the untapered edge. This left 4 pairs of tapered boards, 2 for each leg: one measuring 3½" wide at the untapered end and the other measuring 2¾".

5. **Assemble the legs.**

I brushed on wood glue and tacked each of the 4 leg pairs together with a brad nailer (**Figure D,** previous page). I like to use brad nailers to accurately stitch glued boards into place before adding C-clamps (**Figure E**). This is especially helpful when clamping tapered pieces made slippery with glue.

A meticulous carpenter would use glue blocks to keep the C-clamps from denting the soft pine, but I was racing the setting sun.

6. **Make the tabletop.**

To attach the legs to the tabletop I made a rectangular box out of 1×4. I wanted a lip under my tabletop of (about) 1¾", so I cut 2 long side pieces at 45" and 2 end pieces at 15½". I tacked this box together with the brad nailer, and then, using a straightedge, I scribed a line on the underside of the tabletop, 1½" from each edge. This is one of the beauties of starting with a factory-squared top.

TIP: Pneumatic brad nailers, with fine wire brads, are handy for tacking things into place or stitching multiple pieces together when you don't have a second pair of hands.

Adding glue to the inside of that line, I pressed the rectangle into place, squeezing out the excess glue. Then I grabbed 4 of those scrap pieces I made earlier and screwed them to the underside of the table as temporary mounting blocks to hold the rectangle in place.

TIP: Torx or Star-Drive self-tapping, general-purpose wood screws will spare you the frustration of using Phillips-head screws that tend to "cam out" or strip during hasty projects.

TIP: Self-clamp when you can. Clamping and gluing take time. By using slightly oversized bolts with washers, you can drill, glue, assemble, adjust, tighten, and proceed without waiting.

I added 4 (also temporary) screws to hold the rectangular box to the mounting blocks and the top (**Figure F**).

Then I flipped it over and stitched down the top with sturdy, countersunk, screws. (There are, of course, ways to do this without screwing smack through the top surface of the table if that bothers you. You can buy special hardware for mounting the top from underneath. You can even leave the mounting blocks in place.)

Finally, I flipped it back over to remove the temporary screws and mounting blocks, and I wiped up the extra glue with a damp rag.

7. **Attach the legs and trim.**

From there I went into "Erector set" mode. With the tabletop upside-down on a flat surface, I held each leg in place, drilled 4 holes, slathered on some glue, and then clamped it into place using 4 carriage bolts with washers. If you've cut carefully, the legs will align themselves so that you only need a quick check with a square before wrenching them tight.

Finally, I cut 4 pieces of 1×4 trim to fit tightly between the legs on the outside surface, screwed and glued these into place from the inside (**Figure G**), and I was ready to scrub off all excess glue and sweep the sawdust from my driveway. ◪

Tom Parker (parker@rulesofthumb.org) is an author who lives in Ithaca, N.Y., and works for Cornell University. When he is not tinkering with junk, he's a flight instructor and flies a 1956 Cessna 180 bush plane.

1.2.3.

$5 Smartphone Projector

By *Photojojo/Danny Osterweil* Illustrations by *Julie West*

SLIDE PROJECTORS ARE GREAT, BUT OUTDATED, and digital projectors cost a bundle. Fortunately, you can show off your mobile photos and your phone hack savvy by turning your smartphone into an inexpensive projector.

1. Prepare the projector box.

If the inside of your shoe box is a bright color, paint it black or tape up some black paper for best image quality. >> On a short side of the box, trace the outer edge of your magnifying glass and cut it out. >> For extended use, cut a small hole at the back of your box for your phone's power cord. Tape the magnifying lens securely in place, and make sure there are no holes to let light in.

2. Make a phone stand and flip your screen.

Bend your paperclip into a cellphone stand. >> When light passes though the lens, it gets flipped, so the picture from your projector will come out upside-down. Visit makezine.com/36 for instructions on how to achieve a screen flip.

3. Find your focus.

Project onto a bare white wall or another flat, white surface. >> Position your phone in its stand near the back of the box and walk the box forward or backward until the image starts to come into focus. Fine-tune the focus by moving the phone forward or backward in the box. >> Set your phone's photo app to slideshow mode for a hands-free experience. >> If desired, put the power cord through the hole in the back of the box and seal with a bit of tape. >> For best viewing, turn the screen brightness of your phone all the way up, put on the box top, cover any windows, and turn the room lights down.

Thanks to Instructables user MattBothell for inspiring this project! ↗

Photojojo is on a mission to make photography more fun for everyone! It publishes an insanely great newsletter and carries only the most awesome photo gifts and gear for photographers. photojojo.com

YOU WILL NEED:
» Shoebox » Paper clip » Smartphone » Magnifying glass with low magnifying power » Pencil » X-Acto knife » Electrical or black duct tape » Matte black spray paint or black paper (optional)

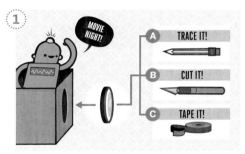

A TRACE IT! | B CUT IT! | C TAPE IT!

PAPER CLIP ORIGAMI!

WHOA!

Vinyl Silk-Screen Printing

⚡ **TIME: 1–2 HOURS**   ⚡ **COST: $10–$30**

Use adhesive "sign vinyl" to silk-screen without the aid of chemicals or a darkroom.

Written by *Chris Connors*

With this technique, you can make your own custom silk-screened T-shirts or posters without any darkroom or chemical access. The one specialized tool you need is a vinyl cutter, but you only need access to it for one step. The vinyl stencil can last for years and allows you to print dozens of images.

1. Prepare the image.

Choose a high-contrast image and crop your selection. Next, convert the image to black and white and if necessary, increase your contrast until you have a fairly simple-looking graphic. (See makezine.com/36 for the Makey the Robot logo and a more in-depth tutorial on creating a graphic.)

2. Ready the file for the vinyl cutter.

Prepare the file with the vinyl cutter's bundled program — I use Roland CutStudio, but you can use Inkscape or another program. Add a box around the image, leaving at least 2"–3" of margin around all sides of your image so that the ink won't seep around the edges.

When you have your design all set, send it to the vinyl cutter. The knife of the cutter will follow the contours of your design.

3. Make a mask for the screen.

If you don't have a silk-screen frame, visit makezine.com/36 for instructions on how to make your own out of a picture frame and

■ **Make:**

MATERIALS

» Plain shirt
» Sign vinyl, self-adhesive
» Screen printing ink, water-based
» Transfer tape
» Masking tape
» Coroplast or MDF

TOOLS

» Photo-editing software such as GIMP
» Utility knife
» Ruler, metal
» Ink knife
» Printing squeegee
» Silkscreen frame
» Vinyl cutter
» Weeding tool

Gunther Kirsch

Tip: If you make the mask by hand, use a straight metal ruler and a sharp utility knife.

silk-screen mesh.

Cut a piece of vinyl the length and width of your screen, plus some extra on all sides, to ensure that it overlaps the frame. In the center of the vinyl, cut a rectangular hole that's big enough for your image. This hole in the mask should be the only place where ink can flow through the screen.

Peel the backing paper from the vinyl and place the mask on the table with its adhesive side facing up. Place the stretched screen frame onto the vinyl, bottom (T-shirt side) down, so it sticks to the adhesive. Make sure the vinyl is attached as neatly as possible to the frame. If your mask is bigger than your frame, make cuts from the edge of the frame to the edge of the vinyl to make it fold neatly.

4. Weed the stencil.

With a weeding tool or utility knife, remove the parts of the stencil where you want the ink to flow through (**Figure 4a**). You can make a more kid-friendly weeding tool by taping a pushpin to a ballpoint pen barrel (after removing the ink cartridge).

Be careful not to remove parts of the image where you want no ink to flow. To keep a complex design clear in your mind as you do the weeding, it's helpful to refer to a black-and-white print of your image.

Weed the whole image and leave it on the backing paper (**Figure 4b**).

5. Add transfer tape to stencil.

Transfer tape is thinner and less adhesive than masking tape. It's designed to hold the vinyl stencil together until you put it on a surface. Choose a roll of tape that's just wide enough to cover your image.

Get a friend's help to make this step easier. One person applies the tape to the vinyl, smoothing it on gradually and evenly. The other person holds the roll with both hands, keeping an even tension on the tape so that it lays down on the vinyl without bubbles or wrinkles.

You can use the squeegee to make sure the transfer tape goes on evenly (**Figure 5a**), but your hands should be enough.

Once you've covered the stencil with the tape, cut it off the roll (**Figure 5b**).

6. Attach stencil to the screen.

Remove the backing paper, making sure the vinyl stencil stays stuck to the transfer tape (**Figure 6a**, following page).

Neatly and evenly apply the stencil to the top of the screen, making sure the image is completely inside the hole in the mask. Burnish or rub the stencil so that it's completely

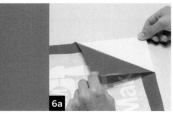
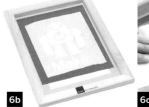

6a

6b

6c

6d

7

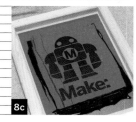

8a

8b

8c

adhered to the screen (**Figure 6b**).

Carefully remove the transfer tape from the stencil, peeling the tape back as flat as possible (**Figure 6c**). If you peel the transfer tape at any angle much less than 180°, the vinyl tends to stay stuck to the tape.

If the stencil doesn't stick to the screen, fold the tape back over, rub the vinyl some more, then try peeling the tape again. Make certain that the stencil stays stuck to the screen.

Fill in any gaps between the stencil and the mask with masking tape (**Figure 6d**). For future projects, when you want to switch to a new stencil, leave the mask on and just peel off the stencil.

7. **Prepare the shirt for printing.**

Remove the tags and stickers from the shirt. You shouldn't need to wash it beforehand (unless it's already dirty). Brand new shirts seem to take the ink just fine.

I like to put a thickness of coroplast or ⅛" MDF inside the shirt. A small stack of newspapers works as well. The insert serves 2 purposes: It keeps the ink from bleeding to the back of the shirt, and it adds a bit of tension to the screen while printing, which helps you make a more crisp print.

8. **Make the print.**

With the ink knife, put some screen printing ink into the well of the frame (**Figure 8a**).

Use the squeegee to draw the ink across your design (**Figure 8b**).

You want to use enough pressure to push the ink through the holes in the screen, but not so much pressure that the ink goes sideways beyond the stencil (**Figure 8c**). This may take a bit of practice. Do some testing to find the right touch for inking the shirts.

When you're done printing, clean the screen promptly with water to remove the ink. You can take the insert out of the shirt, or leave it in during the drying process.

WARNING:
If you skip Step 9, you might ruin a whole load of laundry.

9

9. **Dry and heat set.**

You can speed the drying process up with a hair dryer.

Once your shirt is totally dry, heat-set the ink by putting it in the clothes dryer on high for 20 minutes or so. The label on your ink bottle might specify a more precise time. You can also use a clothes iron to heat-set the ink, but make sure the ink is dry to the touch first. ◪

Chris Connors is a high school art teacher on Martha's Vineyard, who loves to learn with curious people who are interested in inventing the future. He wrote the "Mendocino Motor" project in MAKE Volume 31.

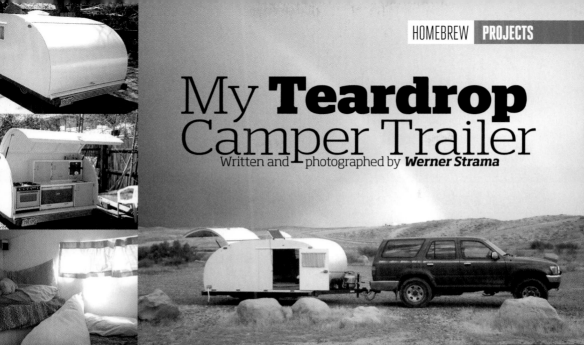

My **Teardrop** Camper Trailer

Written and photographed by *Werner Strama*

⚹ TIME: 300 HOURS   ⚹ COST: $2,000

✚ I got inspired to build my own teardrop camper trailer after a tent camping trip at Lake McConaughy, Neb. The place is beautiful, but the wind was howling, there were monstrous horse flies, frogs all over the place, flying beetles, and at one point the flies and sand made it impossible to cook. I enjoyed it (sort of) but my wife told me that unless we get something better to camp in, she was out.

After searching on the internet I stumbled on a website called Teardrops n Tiny Travel Trailers, and was sold on the small form factor, low weight, and sturdiness of the teardrop shape. I started collecting pictures of all the items that we needed: three-person sleeping area, a small kitchen in the back, self-powered when necessary, and very well insulated, for we would be camping year round. Also I needed to keep the costs down.

I enlisted my neighbor Denny's help, since he's an amazing bargain hunter on camping stuff; he's the one who found the old popup camper for free, as well as the stove with the oven. With the trailer chassis, I calculated dimensions and average weights to maintain 12–15% of the total weight on the hitch, ensuring the trailer would be stable. All it takes is simple math and three bathroom scales. Then it was a matter of getting the wood and extra tools, like a router and a sabre saw.

The actual build time was roughly six months

working on weekends and days off. We set up to make it utilitarian, but not a "work of art," which could take more time. The biggest challenge was assembling the walls, which required lots of woodwork and learning to use the router to cut the channels on the 2×4 wood pieces. Once I learned the tricks, I was able to do flush cuts and shaping in almost no time. I also made a couple of useful jigs to build the cabinets.

The best feature is the size. I can tow it to remote places where an average camper would not get easily. The turning radius is excellent, and a couple of times I was even able to turn around on narrow dirt mountain trails without having to unhook and rotate it manually. Other great features are the insulation and the ceiling fan.

Also, I've been using it as my living quarters when I go contracting out of town, and it has kept me nice and warm in midwinter with outside temperatures of −10°F. I'm planning on covering it in fiberglass, as I've noticed that hail had damaged some of the top. Also, I may remove the propane stove and replace it with a Coleman liquid fuel stove, as propane doesn't work well at altitude or in very cold temperatures. ◪

✚ Make your own: makezine.com/projects/teardrop-camper-trailer.

Werner Strama is an airframe and power plant mechanic based in Colorado.

Kitchen-Table
Cider Making

Written by
**Dr. Nevin
J. Stewart**

⚡ **TIME:** 2 HOURS, 2 PEOPLE ⚡ **COST:** $75–$150

Forget messy presses — use modern centrifugal juicers.

MATERIALS

FOR 5 GALLONS OF CIDER:
» **Apples, twice washed, 75lb**
» **Yeast, champagne variety,
  Saccharomyces bayanus, 5g**
» **Campden tablets** for sanitizing
» **Sugar (optional)** for sparkling cider

TOOLS

*Specialty brewing items are available
at homebrewing and winemaking stores.*

FOR BREWING:
» **Juicer, whole fruit, with spout** such as
  the Breville Juice Fountain Elite (or Plus).
  Buy the highest power rated machine
  your budget will permit. Try secondhand
  shops, yard and garage sales, and eBay.
  I've yet to pay more than $45.
» **Food-safe plastic hose** to fit juicer
  spout; typically 1" ID × 16" long
» **Cotton towels**
» **Spring clamp**
» **Fine straining bag** typically 24"×24"
» **Brewing bucket, 5gal, open top, with
  tap at bottom** typically 12" diameter ×
  17½" tall
» **Pails, 2gal (2)** for sterilizing and strain-
  ing. The straining pail should be the
  diameter of your brew bucket, or a bit
  larger; drill 30+ holes in the bottom.
» **Hydrometer (optional)**
» **Funnel**
» **Carboy, 5gal (1) or demijohns, 1gal (5)**
» **Airlock(s) and rubber stopper(s)** adds
  about 5" to your demijohn/bucket height
» **Measuring cup, 2c**

FOR RACKING AND BOTTLING:
» **Rubber tubing** for siphoning
» **Siphon tap (optional)**
» **Tray with rim** such as a cookie sheet
» **Beer bottles**
» **Bottle caps, crown style**
» **Hand-operated bottle capper**
» **Labels**

A

➕ Yes, our new neighbor got as drunk as a
skunk. Blootered, bladdered, and blitzed
as we Scots say or, in Cockney slang, Brahms
and Liszt! He and his partner came along to
a cider-themed evening at our house where
he was initiated into the Scillonian Road hard
cider making cooperative.

We started the evening with a hands-on practical
session of how to use my "juice and strain" method
to make clear apple juice and crystal-clear hard cider,
quickly and with minimal mess.

Dinner was accompanied by last year's homemade
hard cider (6.5% alcohol by volume), apple wine (15%
ABV), and store-bought Calvados (40% ABV). It was
that last beverage that did him the most damage.

The next day our new friend could remember nothing
of the evening's proceedings. He was unable to recall
disclosing his lifetime's accumulated prejudices
concerning ethnic minorities, politics, religion, and

relationships. Also my introduction to the juice and strain cider method had been lost, as with the spirit vapors.

To save further embarrassment, I said nothing about his verbal indiscretions, but I did re-explain my method to him as follows.

1. Clean your equipment.

Apple juice and hard cider are foodstuffs, and all appropriate food handling and safety measures should be followed. Wash your hands, sanitize all surfaces, double-wash the apples, and throw away any bad ones.

Sterilize all equipment that will be in contact with fresh apple juice. I use a stock solution of 4 Campden tablets per gallon of water to soak all the relevant parts and buckets for a couple of hours before use.

2. Set up the juicer and strainer.

Lay out a clean towel, rinse off the juicer parts, and assemble your whole fruit juicer.

Attach the "juice containment and delivery adaptor," aka hose, to the juicer's spout, and feed it into the straining bag that is held within a straining pail with holes in its base. This assembly sits neatly in the open brewing bucket with a draw-off tap at the bottom.

Set up your brew bucket on a stool or box, high enough that you can fit your demijohn or carboy underneath the tap. Apples go in at one end, clear apple juice comes out at the other. It couldn't be simpler (**Figure A**).

3. Juice and strain.

Feed apples into the juicer with a steady, even pressure on the pusher. The higher the machine's power rating, the faster you can go (**Figure B**).

When the pulp container fills up, discard the pulp. After every 25lbs of fruit, dismantle the machine and clean the pulp off the centrifuge stainless steel mesh (**Figure C**).

You'll find that the juicing work is done in a flash, although it takes a while longer for all the juice to strain through. I obtain the last 5% of the expected 65% by weight of juice by wringing out the straining bag. Scottish, you see!

CIDERMAN, CIDERMAN: Homebrewers Nick McDuff and Nevin Stewart present the coveted Onslow's Dry.

IMPORTANT: If you're making only fresh apple juice, use only handpicked apples, as windfalls may have harmful bacteria which might not wash off.

CAUTION: Whole-fruit juicers are powerful appliances. Read and adhere to the safety instructions for your juicer.

F

G

H

+ Recommended reading: *Real Cidermaking on a Small Scale* by Michael Pooley and John Lomax; *Craft Cider Making* by Andrew Lea

What's left in the bag is about 1% of the original apples. This very fine pulp can be used in apple muffins. You don't want it in your cider.

4. "Shoot" the yeast.

While the last juice is draining, pitch the yeast into a measuring cup containing fresh, clear apple juice held at room temperature (**Figure D, previous page**). This will allow the dried yeast to rehydrate and kick-start your fermentation. Use a champagne yeast for simplicity and reliability. A 5g packet is enough to inoculate 5gal of juice.

At this point, measure the original gravity (OG) with a hydrometer and write it down (**Figure E, previous page**). Later, this figure will allow you to estimate your cider's alcohol percentage. If the OG is low, top it up with a little white sugar to reach 1.040.

After half an hour, stir the cup to thoroughly disperse the yeast, then pour it into your sterilized carboy or demijohns (**Figure F**). Fill these up nearly to the top with juice and put airlocks on top (**Figure G**). Within the hour you should see bubbles coming out through the airlock.

5. Ferment.

Keep the fermentation vessel(s) in a warm place like the kitchen and after 3 weeks you should have a crystal-clear cider ready to be racked and bottled. Check it with your hydrometer. The reading needs to be 1.000 or less. If it's still high, let fermentation continue.

When the hard cider is finished, measure the final gravity and read off the alcohol content from an ABV chart or online calculator. For reasonably good storage, 5% ABV is considered the target minimum.

6. Bottle.

Siphon your cider into recycled, sterilized beer bottles that will take a crown cap (**Figure H**). If you want a still hard cider, just bottle as is, and label (**Figure I**).

If you want bubbles, then add ½ teaspoon of white sugar to a pint bottle, fill up with your hard cider, and cap. After a few more weeks, a secondary fermentation should be complete and you'll have some fizz.

7. Enjoy.

Serve chilled. Take care when opening. If you've overdone the sugar, it can go off like a fire extinguisher. You can adjust the sweetness to taste by adding sugar syrup. But I prefer my cider as dry as it can go. ◪

Dr. Nevin J. Stewart is a Presbyterian, husband, father, cook, food shopper, Scot, chemist, inventor, cider maker, gardener, and country walker. He lives in Guildford, an English cathedral and university town not yet granted city status, and his favorite wine is pinot noir.

I

Gregory Hayes

Lite-Brite
LED Clock

Written by
Lonnie
Honeycutt

⚡ **TIME: 6 HOURS**   ⚡ **COST: $40–$50**

Swap passive pegs for programmable light-emitting diodes and build a one-of-a-kind, Arduino-powered desk clock.

I had just finished mentoring a high school student through the build of a 4×4×4 LED cube. I was sitting at my desk, looking at a small pile of leftover LEDs, trying to think of something fun to do with them. That's when my 3-year-old daughter walked into the room, sat down, and started playing with her little pink Lite-Brite. I watched her put a peg into the screen and was struck by how much it looked like a little

LED, glowing there. This was all the inspiration I needed! Check out my step-by-step build instructions online at makezine.com/36. ↗

Lonnie Honeycutt is a married father of three children, who first learned electronics as an Air Force radio systems technician. He got into hobby projects when he discovered Arduino. Lonnie works as a poker dealer in Baton Rouge, La., and writes about projects and electronics in general at meanpc.com.

Country **Scientist**

⌁ **TIME: 1 HOUR**   ⌁ **COST: $5–$10**

How to Use LEDs to Detect Light

Written and photographed by *Forrest M. Mims III*

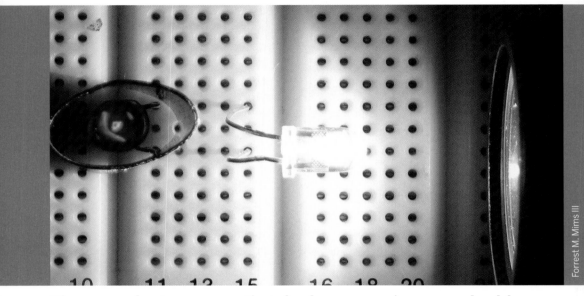

Forrest M. Mims III

Since an electromagnetic telephone receiver can double as a microphone, can a semiconductor light detector double as a light emitter?

That question was on my mind when I was a high school senior in 1962. Back then I didn't know that quantum effects in a semiconductor are unrelated to the electromagnetic operation of a telephone receiver. If I'd known that, I never would have connected a spark coil across the leads of a cadmium sulfide photoresistor to see if it would emit light. It did — a soft green glow punctuated with bright flashes of green.

During college I found that a silicon solar cell connected to a transistor pulse generator emitted flashes of invisible infrared that could be detected by a second solar cell. In 1972, I used near-IR LEDs and laser diodes to send and receive voice signals, through the air and through optical fibers. Later I experimented with two-way optocouplers made by taping a pair of LEDs together so they faced one another.

In 1988, I tried LEDs as sunlight detectors. They worked so well that the first homemade LED sun photometer I began using on Feb. 5, 1990, is still in use today.

Why Use LEDs As Sensors?

Silicon photodiodes are widely available and inexpensive. So why use LEDs as light sensors?

» LEDs detect a narrow band of wavelengths, which is why I call them *spectrally selective photodiodes*. A silicon photodiode has a very broad spectral response, about 400nm (violet) to 1,000nm (invisible near-IR), and requires an expensive filter for detecting a specific wavelength.

» The sensitivity of most LEDs is very stable over time. So are silicon photodiodes — but filters have limited life.

» LEDs can both emit and detect light. This means an optical data link can be established with only a single LED at each end, since separate transmitting and receiving LEDs aren't needed.

» LEDs are even more inexpensive and widely available than photodiodes.

Drawbacks of LEDs As Light Sensors
No sensor is perfect.

» LEDs are not as sensitive to light as most silicon photodiodes.

» LEDs are sensitive to temperature. This can pose a problem for outdoor sensors. One solution is to mount a temperature sensor close to the LED so a correction signal can be applied in real time or when the data are processed.

» Some LEDs I've tested do gradually lose their sensitivity.

LEDs Detect Specific Colors of Light
The typical human eye responds to light with wavelengths from around 400nm (violet) to about 700nm (red). LEDs detect a much narrower band of light, having a peak sensitivity at a wavelength slightly shorter than the peak wavelength they emit. For example, an LED with a peak emission in the red at 660nm responds best to orange light at 610nm.

The spectral width of light emitted by typical blue, green, and red LEDs ranges from about 10nm–25nm. Near-IR LEDs have a spectral width of 100nm or more. The sensi-

NOTE: This does not apply to white LEDs, which are blue-emitting LEDs coated with a phosphor that glows yellow and red when stimulated by blue from the LED. The merging of the blue, yellow, and red provides white light. While a white LED can detect blue light, a blue LED is a much better choice.

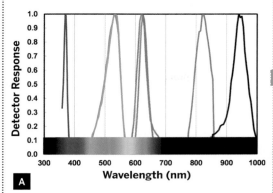

A

tivity of most LEDs I've tested provides ample overlap to detect light from an identical LED.

Figure A shows the spectral response of 7 blue, green, red, and near-infrared LEDs that replace the usual silicon photodiodes and filters in my modified Multi-Filter Rotating Shadowband Radiometer, used for solar spectroscopy.

Blue and most green LEDs are made from gallium nitride (GaN). The brightest red LEDs are made from aluminum gallium arsenide (AlGaAs). The LEDs used in near-infrared remote controllers are also AlGaAs devices; their peak emission is about 880nm and peak detection around 820nm.

Older remote controllers used gallium arsenide compensated with silicon (GaAs:Si). These LEDs emit at about 940nm, which makes them ideal for detecting water vapor, but they've become very difficult to find.

In my experience, the sensitivity of red "super-bright" and AlGaAs LEDs and similar near-IR LEDS is very stable over many years of use. Green LEDS made from gallium phosphide (GaP) are also very stable. However, a blue LED made from GaN has declined in sensitivity more than any LED I have used.

Basic LED Sensor Circuits
You can substitute an LED for a standard silicon photodiode in most circuits. Just be sure to observe polarity. Also, remember that the LED isn't as sensitive as most standard photodiodes and will respond to a much narrower band of light wavelengths.

For best results, use LEDs encapsulated in clear epoxy and try a few experiments first. These will help you understand how the detection angle of an LED used as a sensor matches its emission angle when used as a light source:

» Use standard couplers to attach LEDs to plastic optical fiber, or attach them directly using this method (**Figure B**): Flatten the top of the LED with a file, clamp it securely, and carefully bore a small hole just above the light-emitting chip. Insert the fiber and cement it in place.

» Connect the leads of a clear encapsulated red or near-IR LED to a multimeter set to indicate current. Point the LED toward the sun or a bright incandescent light, and the meter will indicate a current (**Figure C**).

» Use one LED to power a second LED. Connect the anode and cathode leads of 2 clear encapsulated super-bright red LEDs. When one LED is illuminated with a bright flashlight, the second LED will glow. Heatshrink tubing is placed over the glowing LED to block light from the flashlight. You can see this working in the photo on page 136.

LEDs have a much smaller light-sensitive surface than most silicon photodiodes, so they're more likely to require amplification. Inexpensive operational amplifiers are ideal. **Figure D** shows a simple circuit I often use to convert the photocurrent from an LED into a proportional voltage. The Linear Technology LT1006 single-supply op-amp (IC1) provides a voltage output that's almost perfectly linear with respect to the intensity of the incoming light. The gain or amplification equals the resistance of the feedback resistor (R1). Thus, when R1 is 1,000,000 ohms, the gain of the circuit is 1,000,000. Capacitor C1 prevents oscillation.

Many other op-amps can be substituted for the LT1006, but most of them require a dual-polarity power supply. If you use one of these, connect pin 4 directly to the negative supply. Connect pin 3 and the cathode of the LED to ground (the junction between the minus side of the positive supply and the positive side of the minus supply).

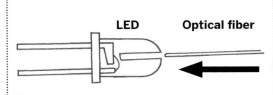

LED   **Optical fiber**

B

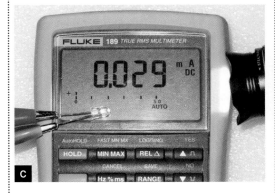

C

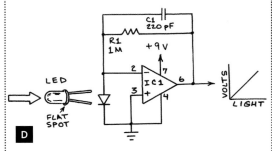

D

Going Further

The best way to come up with new applications for LEDs operated as photodiodes is to experiment with the applications I've described here. When I was doing this back in the 60s, I had no idea those simple experiments would lead to two-way communication over a single optical fiber and several kinds of instruments to measure the atmosphere that I've been using for more than 23 years. ▨

✚ Some of my articles and books that describe applications for LEDs as photodiodes are listed at forrestmims.org/publications.html.

Forrest M. Mims III (forrestmims.org), an amateur scientist and Rolex Award winner, was named by *Discover* magazine as one of the "50 Best Brains in Science." His books have sold more than 7 million copies.

TOY INVENTOR'S NOTEBOOK

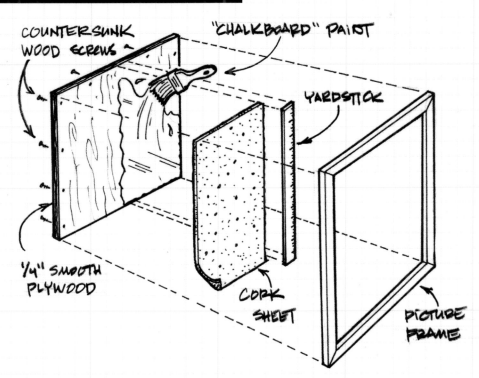

COUNTERSUNK WOOD SCREWS

"CHALKBOARD" PAINT

YARDSTICK

¼" SMOOTH PLYWOOD

CORK SHEET

PICTURE FRAME

"Old School" Inventing

On a recent family trip, I was asked to help out on a quick DIY project. My sister-in-law Carole was making a combination blackboard/bulletin board from some wooden picture frame material and a piece of ¼" smooth plywood for the back panel.

She painted one half of the panel with chalkboard paint. Just brush it to get an "old school" chalkboard surface – works great! She wanted some help with the next step; I attached pieces of cork sheet with contact cement for the bulletin board side.

The last detail was trickier: How to protect the exposed tender edge of the cork? If I were at home in my shop, I'd use a router to make a wooden trim strip with an L-channel shape, but away on vacation I had to improvise. Hmm…old school…classroom…rulers? Aha! I trimmed an old wooden yardstick to fit inside the frame and abut the cork edge. Easy, fast, and free from the hardware store. The ruler markings and faded red lettering on the yardstick even adds a fun "old school" look. Project's done and school's out – back to my summer vacation! ◪

> ➕ See photos of the finished bulletin board at makezine.com/36.

Hedge Maze
Area Rug

Give your rug a haircut.

Written by *Sean Michael Ragan*

Gregory Hayes

⟋ TIME: 6 HOURS   ⟋ COST: $30–$60

MATERIALS

» **Carpet, ¾" pile, sized for your design** I used an inexpensive synthetic from a large chain hardware store.
» **Masking tape** of about the same width as your clipper head. I used Frog Tape #82031, 1⅞" × 60yds.
» **Carpet tape** (optional)

TOOLS

» **Computer or pen and paper**
» **Hobby knife** with fresh blades
» **Hair clipper, electric, wall-powered** Battery-powered clippers will probably be underpowered. I used Wahl's "Designer" model #8355.
» **Carpet shears or utility knife**
» **Broom and dustpan**
» **Vacuum cleaner**

Sean Ragan

1

2

✚ I once saw a pricey designer "labyrinth" carpet in a catalog and wondered if I could re-create the effect cheaply by taking electric hair clippers to a piece of ordinary carpet. Long story short: it works. A maze pattern on green carpet is great for the "hedge maze" look, but your design could be anything!

1. Measure, measure, measure.

Measure your carpet section (mine was 74"×71"), the area you want the rug to fill (74"×41"), the width of your shaver head (1.75"), and the width of your masking tape (1.875"). Write it all down someplace.

2. Design the maze.

2a. Verify that you have enough carpet to make the rug you want, and that your masking tape is at least as wide as the head of your clipper. They don't have to be exactly the same, but they should be within ¼" of each other.

2b. Divide the length and width of the area you want to cover by the width of your tape and round to the nearest odd integer. In my case:

$$x = 74" / 1.875" = 39.47 \ (39)$$
$$y = 41" / 1.875" = 21.87 \ (21)$$

2c. Work out a maze on a square grid that is x by y units (39×21, in my case). You can design on paper or in software, or you can generate the maze procedurally. I used John Lauro's simple web-based maze maker at makezine.com/go/mazegen.

3. Lay out the grid.

3a. Starting from the best corner of your carpet, adhere a strip of tape all the way along one edge. If one side of your design is longer than the other, lay out the longer dimension first.

3b. Apply a short "spacer" strip beside the full strip at each end, to make sure you're spacing the corridors consistently, then add another full strip. Keep alternating full strips and spacers until the total number of strips equals the number of units in your maze's short dimension.

3c. Repeat Step 3b along the other dimension, starting from the same corner, until you have a complete grid covering the full area of your design.

4. Cut out the pattern.

Using a new, sharp hobby knife blade, cut out individual squares from the grid to form the corridors in your maze plan. Peel up the cutouts with your fingers and discard them.

5. Clip the corridors.

Plug in your clippers and trim the carpet about ⅜" shorter in the corridor areas.

Watch the length of the carpet piles coming off in front of the blade to monitor the depth of the cut. It doesn't have to be too precise. You can try using guide combs on your clippers, but I found them more trouble than they were worth.

Oil your clipper blades frequently, and take a break now and again to let them cool off.

6. Cut to shape.

Following the outside edge of the tape, cut the perimeter of the rug to shape with carpet shears or a utility knife.

7. Clean up.

7a. Remove the tape. Just grab and pull.

7b. Inspect the corridors for shallow areas, bumps, or other imperfections. Touch them up with the clippers as needed.

7c. Pick up the rug, shake it out hard, and sweep up the loose trimmings. Give it a good vacuuming, and you're done.

3b

3c

4

NOTE: If all your walls and corridors are 1 unit wide, you can lay out your maze by cutting single squares from this grid. Otherwise, you can add short tape sections for "off grid" walls.

TIPS: Practice on a piece of scrap carpet, and work on a smooth floor that's easy to clean up.
No downward force is needed besides the clipper's weight.

5

6

7

NOTE: The carpet backing may be visible at the edges at first, but a few weeks' use will round them over.

TIP: If you're installing the rug on a smooth floor, use carpet tape to prevent sliding.

Going Further

The tedious bit was not the trimming, but applying the pattern. I considered mounting a projector on the ceiling so I could project the pattern onto the carpet, but the tape trick won out for simplicity and cost. There must be better ways to solve the pattern problem: Chalk? Washable paint? Freehand cutting? Let us know at makezine.com/go/hedgemaze! ◪

Sean Michael Ragan is technical editor of MAKE magazine. His work has appeared in *ReadyMade*, *c't – Magazin für Computertechnik*, and *The Wall Street Journal*.

Squire Whipple

and the

Truss Bridge

**Build the simple span
that launched the
great age of iron
bridge building.**

Written by *William Gurstelle*

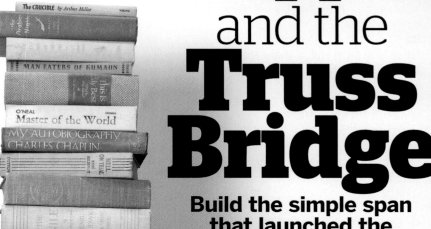

Gregory Hayes

⟋ **TIME: 1–2 HOURS**   ⟋ **COST: $10–$15**

✚ Like all mechanical, civil, and aerospace engineers, I learned
how to analyze structures early in my engineering education.
Determining the size and direction of compressive or tensile forces
acting on each piece of a construction is called *statics*. With this
basic bit of knowledge, we can figure out the right sizes, shapes, and
thicknesses to use in building things.

In statics, we learn about architectural elements such as arches, beams, buttresses, trusses,
and vaults. The truss is one of the most important, and it's widely used in construction. In its
simplest form, a truss is a rigid framework of bolted-together triangles. Triangles are inherently
strong and stable, and structures made out of them are also strong, rigid, and lightweight.

MATERIALS

» Craft sticks or tongue depressors, 6"×11⁄16" (100)
» Bricks, standard, 2¼"×3¾"×8" (2)
» Balsa wood sheet, 1⁄8"×4"×16"

TOOLS

» Utility knife
» Ruler
» Hot glue gun and glue
» Plate weights or concrete blocks (optional) for testing structure

BROOKLYN BRIDGE

James Burke

Who first figured out the truss? It's clear that the classical Greeks, for all their genius, knew little or nothing about building with triangle trusses. The Romans dabbled, but examples of Roman structures that use trusses are few and far between.

Medieval cathedral and church architects understood the technique empirically, if not scientifically; in plenty of early European buildings, wooden triangular trusses hold up the roof. Still, the best-known use of trusses is in building bridges.

For that, we can thank New York civil engineer Squire Whipple (1804–1888), who developed the first scientific method for analyzing and designing trusses. In 1847, Whipple published *A Work on Bridge Building*, which revolutionized civil engineering. No longer would builders use "rules of thumb" to guess at how big to make a strut or girder. Because of Whipple's work, they knew *exactly*. Whipple figured out how to analyze trusses with a graphical method of lining up force vectors that he called "the polygon of forces." This allowed engineers to safely and economically design viable truss bridges without knowing even a bit of calculus or trigonometry.

Whipple's Bowstring Arch bridge, made with cast-iron trusses, became the standard for bridges over the Erie Canal. For his contributions, the Society of American Civil Engineers declared Whipple the "father of American iron bridge building."

Whipple's book set off a boom in civil engineering. Soon, the great age of iron bridge building was in full swing. The triangle structure is obvious in the beautiful spans of the Navajo Bridge over the Colorado River, the Whirlpool Rapids Bridge over Niagara Falls, and the Quebec Bridge over the St. Lawrence River. But hidden trusses are also important parts of suspension and cantilevered bridges like the Brooklyn Bridge.

QUEBEC BRIDGE

WHIRLPOOL BRIDGE

The Warren Truss

The simplest truss bridges are the Howe, the Pratt, and the Warren, named for the engineers who devised them. In 1848, English engineer James Warren designed the first truss bridge consisting solely of equilateral triangles connected side by side, each triangle inverted in relation to the next. Structural engineering doesn't get much simpler. The Warren truss is tried and true, providing a bridge that's easy to build, strong, and relatively lightweight. In 1850, Squire Whipple designed the first Warren truss bridge in America.

In this project, we'll make a model Warren truss bridge. Light and strong, it can hold more than 100 pounds.

NAVAJO BRIDGE

Build a Warren Truss Bridge

1. Cut the gusset plates.

Use the utility knife to cut the balsa wood into 14 squares measuring 2" on a side (**Figure A**).

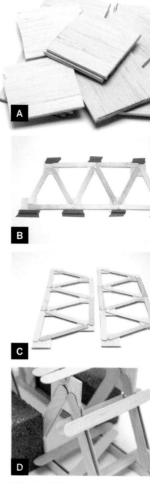

A

2. Build 2 trusses.

Begin by taping 7 gusset plates to your work surface as shown. Hot-glue the craft sticks to the gussets. Take care to make the glued connections neat, aligning the craft sticks to form tidy equilateral triangles (**Figure B**).

Once the glue sets, flip over the truss and attach craft sticks to the other side in the same fashion, for double strength.

Build a second truss in the same way (**Figure C**).

The long top and bottom members of the truss are called *chords*. The slanted members that tie the chords together are *webs*. If you use Squire Whipple's polygon of forces to analyze each member, you'll find that the greatest forces in your truss are in the top and bottom chords, at the center of the bridge. If you wish, attach an extra craft stick at those points for reinforcement.

3. Add struts and braces.

Place the bricks on your work surface 4" apart and make certain the long sides are parallel. Stand each truss vertically against a brick, and tape 3 of its web members to the brick.

Then glue the *struts* and *braces* across the top and bottom, as shown in **Figures D** and **E**. (Struts are braces that meet the chords at 90° angles.) It's essential that the trusses are exactly vertical, perfectly perpendicular to the struts and braces. Even a small amount of leaning will cause your bridge to fail prematurely.

Finally, add braces across either end of the bridge (**Figure D**).

Howe Truss

Pratt Truss

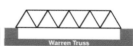

Warren Truss

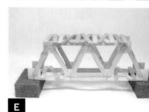

D

E

Test Your Truss Bridge

1. Place the bricks 14" apart on the floor.
2. Place the ends of your bridge on the bricks (**Figure E**).
3. Load 'er up! You can use concrete blocks, barbell weights, buckets of water, or anything else flat and heavy (**Figure F**). Add weight slowly and incrementally, and keep fingers and toes away from the area underneath the bridge. Distributing the weight evenly will allow you to add more weight before the structure fails.
4. It's your bridge, so you can either add weight until it eventually fails, or paint and display it as proof of your engineering abilities. ◹

F

William Gurstelle is a contributing editor of MAKE. The new and expanded edition of his book *Backyard Ballistics* is now available in the Maker Shed (makershed.com).

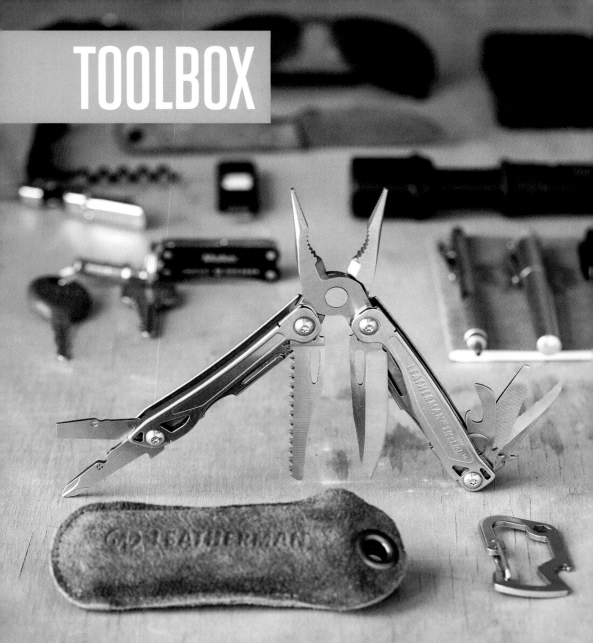

Leatherman Sidekick $43 leatherman.com

»I've carried the same Leatherman Super Tool for 19 years, and it's still going strong. I recently decided to update my everyday carry. After trying several multi-tools, I discovered the Sidekick. It's solid, handsome, and opens to deploy 14 tools (including two pliers, wire cutters, two knives, file, saw, wire stripper, screwdrivers) that combine most of my favorite features from across the Leatherman line.

–Gregory Hayes

Gregory Hayes

READER FAVE

Milwaukee Tools 6519-31 Sawzall Recip Saw Kit

$125 *milwaukeetool.com*

Milwaukee has a reputation for tough, heavy-duty tools that last a lifetime; the Sawzall my dad bought more than 20 years ago is still running strong. We've used it to fix cars, finish basements, remodel houses, build half-pipes, and lots more. Compared to my heirloom tool, there are notable improvements: Instead of using a hex wrench to change the blades, for instance, this model features Milwaukee's "Quik-Lok" blade clamp that requires no tools to operate.

This one clocks in at 12 amps, but if you want a little more power, there's also a 15-amp "super" model that also adds a keyless adjustable foot plate for angled cuts. The Sawzall now comes in a plastic, injection-molded tote that carries saw, blades, and a few odds and ends. Probably works better as a case than Milwaukee's classic red metal toolbox, honestly, but it does not have the same cool style. —*Jake Spurlock*

Padcaster

$149 *thepadcaster.com*

Turn your iPad into a professional moviemaking rig! The Padcaster consists of a tough aluminum frame with a urethane insert to securely hold your full-size 10" iPad (version 2 or later). Besides protecting the tablet, the frame is studded with two dozen ¼" and ⅜" threaded holes for attaching mics, lights, power cables, tripods, and other video accessories. Comes with a bundled shoulder strap and mounting lugs for easy carry and stabilization during use and, even cooler, a 72mm to 58mm lens adapter so you can attach a more traditional DSLR lens.
—*John Baichtal*

Extech BR250 Video Borescope

$350 **extech.com**

A borescope is one of those tools that's probably not in your hand-me-down toolbox. But once you've used one to inspect hard-to-reach spaces, you'll wonder how you ever got by without. This model has a detachable viewer that lets you run the scope up to 32' from the screen, a huge plus as it can turn a two-person job into solo work. Even better, the viewer can record still images and video to an onboard SD card for later review, and comes with magnetic mounting feet and an RCA video out port.

The camera's long 36" gooseneck is fully waterproof and only slightly wider than a pencil, so it's easy to maneuver even in very tight spaces. It has just the right balance of rigidity and flexibility, holding whatever shape without dipping as you move it around, and comes with magnetic pickup and inspection mirror attachments for the tip.

—*David Cook*

Weller Red Standard Duty 25W Soldering Iron

$20 **apexhandtools.com/weller**

In the MAKE Labs, we have more soldering irons than desks. For bench work we use variable-power irons with temperature control stations, but for "field" projects — like the wiring in our Power Wheels go-kart — we love Weller's fixed-power "Red" series. Their 25-watt "Standard Duty" model is our outdoor go-to and "rough duty" tool, and when facing tight spaces, tiny part labels, or dark workplaces, the built-in LED illuminator can be a lifesaver. It wouldn't be my first choice for tacking down SMD chips, but a temperature-controlled bench iron isn't exactly ideal for repairing wire in the crawl space either. New tips are as common as burger joints, and it's nice to be able to see my work in the dark, for a change. —*Sam Freeman*

PanaVise PortaGrip Universal Cellphone Holders with Suction Mounts

$30-$60 panavise.com

Take your favorite workholding accessories on the road with the PanaVise PortaGrip cellphone holder line. Fans will instantly recognize the PortaGrip's styling, reliability, and flexibility as PanaVise hallmarks. These two models feature foam-backed, fully adjustable tilt cradles that allow you to position and rotate the phone as you like. The padded bumpers are also adjustable, ratcheting along the bottom and sides as needed to allow access to your phone's ports and buttons, and their low profiles won't interfere with your touchscreen. The shorter reach of the 811 is perfect for the closer windshield of my Jeep, and the 709B's long arm works great in the larger cab of the MAKE Mobile fire truck.　　—GH

Gregory Hayes

BriskHeat Plastic Bending Strip Heater Element

$79 tapplastics.com

If you're interested in bending sheet plastic — say, for instance, because you were inspired by Charles Platt's " Fantastic Plastic Desk Set" article in MAKE Volume 10 (page 100) — you'll get best results with a strip heater, which is pretty much just what it sounds like: a long, straight heating cord that softens the plastic in a narrow line along which it then easily and cleanly bends. Pro-grade units can be a bit pricey, but it's easy to achieve comparable results with a homemade unit that costs about half as much to build.

One thing you will probably want to buy for your homebrew bender is the heating element itself, which is where TAP Plastics' BriskHeat comes in. The BriskHeat consists of a ½" x36" heating element covered in a heat-resistant weave; the electrical plug is split, with the power lead at one end of the strip and ground at the other, which avoids having to use pricey heatproof insulation on the return wire. The BriskHeat can bend plastics up to ¼" thick, quickly and cleanly — a fact anyone who's ever tried doing it with a heat gun will appreciate.

　　—JB

3M Scotchlok IDC Butt Connectors

$5/25 solutions.3m.com

3M's Scotchlok quick connectors are incredibly handy for loose wiring jobs. I tested their UY2 "butt" connectors, intended to join two conductors end-to-end, but the series also includes cap, tap, and even four-way connectors in various gauges. To use, just insert the cut ends of the wires, aligning them with the internal metal slot terminals, and push down on the tab to make the connections. 3M only warrants their published specs when the deal is sealed using their specially-designed E9-Y hand crimping tool, but, between you and me, we had pretty good luck with regular pliers, too. Much faster, better looking, and (I expect) more durable than wire nuts. —*Kelley Benck*

Gregory Hayes

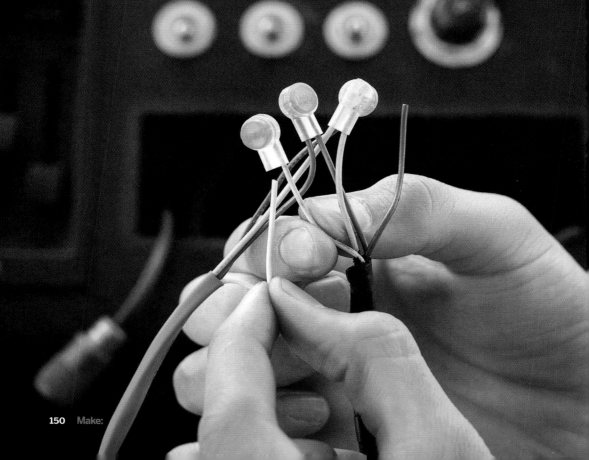

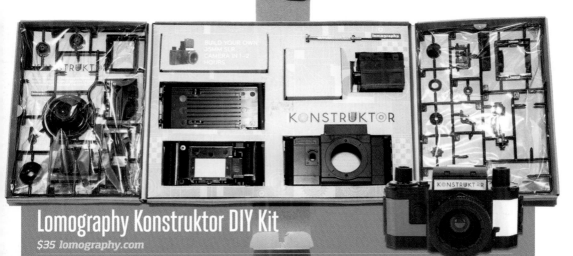

Lomography Konstruktor DIY Kit

$35 lomography.com

The packaging on this kit is the best I've ever seen. The box is beautiful, and the parts are laid out so well I almost felt bad taking them out. The instructions are a bit daunting, but the videos are clear, and between the two, I was able to complete the "one-to-two hour" build in about an hour and twenty minutes.

Shooting is a unique experience. The old-school ground glass makes for a vintage feel; there's no prism, so images are left/right reversed or upside down. The shutter and viewfinder are connected, which feels strange, but makes taking gorgeous double-exposures easy. Focusing is rough, aperture is stuck at f/10, and the shutter is either 1/80 or bulb. Advancing is weird, so I ended up with wide gaps between some frames and some photos overlapping.

But that's the idea: The rough bits are part of the charm. And there wasn't a single light leak on any of my prints. Get this for anyone looking for a fun kit experience, especially Instagram enthusiasts who may appreciate being reminded what their favorite filters are based on.   *—SF*

Pelican 35QT Elite Cooler

$200 pelican.com

Want to keep your food cold during an afternoon hackathon? Grab a styrofoam tub and some ice. But if you need a cooler that takes a lickin' and keeps on tickin', check out the Pelican ProGear Elite series. I played around with their 35-quart model, the smallest of the line, which ranges up to a monstrous 250-quart chest. They all pack the same heavy-duty construction: 2" polyurethane insulation and a freezer-grade gasket, boasting ice retention of up to 10 days. I left mine on the deck one Sunday, sitting in the hot sun and opened at least once a day, and there were still chunks of ice inside on Friday. That performance has a price, however: The thing's a tank, weighing in at 32 pounds bone dry. Still, the cooler's efficiency, durability, and feature-packed design (the drain plug is threaded for a garden hose, for instance) make it totally worth its weight, especially in those situations where performance counts more than portability.

—JB

BOOKS

New from MAKE

Zero to Maker
by David Lang

$20 **Maker Media**

The essential guide to joining the maker community, finding resources, learning new tools and skills, and developing a new maker lifestyle, career, or both.

The Makerspace Workbench

$30 **Maker Media**

Complete guidance on setting up and running a makerspace environment.

Getting Started with Sensors
by Tero & Kimmo Karvinen

$17 **Maker Media**

Everything beginners and experienced users need to know to create sensor-based projects with Arduino and Raspberry Pi.

Things Come Apart by Todd McLellan

$30 **Thames & Hudson**

Since this book arrived in our office, it's been hard to hold on to long enough to even flip through all the pages. Canadian photographer Todd McLellan has elevated the teardown to the status of high art, and we dig his work so much here at MAKE that we've included our own little homage, in this issue, on page 120.

McLellan's art involves meticulously disassembling manufactured objects and photographing their parts with two opposing processes. In the "portrait" process, the bits are carefully arranged on a flat, neutral background and shot from directly overhead. In the "candid" process, it's all just dumped in a jumble and captured mid-fall using a strobe — the result is something like a real-life exploded diagram.

The simplest of the book's subjects (a mechanical pencil) consists of just 16 components, and the most complex (a light kit aircraft) more than 7,000. Forced to pick a favorite, I'd probably opt for the gorgeous four-page foldout that opens to reveal a painstakingly dissected 1912 Gerhard Heintzman upright piano. The time-series on the outside folds — depicting, I believe, McLellan's efforts to drop all the piano's pieces at once — is pretty entertaining, too.

It's an oversized book, at 10"×14", and unlike a lot of the review copies that come across my desk, probably worth owning as an actual book; no tablet I know of could really do it justice.

Though the photos take center stage, *Things Come Apart* also includes some insightful prose, both in McLellan's foreword and in a series of interstitial essays by notables like iFixit CEO Kyle Wiens and art conservator Penny Bendall. Here, too, the tension between order and chaos is a recurring theme: Wiens, for instance, waxes convincingly poetical about repair as a deep metaphor for life itself — after all, it's fundamentally about overcoming entropy, right? Even the book's title seems chosen as a clever play on the famous line from Yeats' *The Second Coming* and, indeed, to hint at a more philosophical response.

—Sean Michael Ragan

The Good Life Lab: Radical Experiments in Hands-On Living

by Wendy Jehanara Tremayne

$19 Storey Publishing

Several years ago, makers Wendy Tremayne and Mikey Sklar decided to leave their high-pressure jobs in New York City and move to rural New Mexico, where they "made, built, invented, foraged, and grew all they needed to live self-sufficiently." The incredible homestead they created using materials from the waste stream is testament that resourcefulness and a willingness to learn can make anything possible. This book chronicles their path to a decommodified maker lifestyle, their journey to learn how to build and make the things they need, and the abundant life they lead as a result. Filled with how-tos on everything from building with papercrete to hacking appliances for maximum efficiency to making your own mead, it's 300 beautifully bound pages of inspiration. Read it and get ready to quit your day job.

—*Goli Mohammadi*

Arduino Workshop: A Hands-On Introduction with 65 Projects

by John Boxall

$30 No Starch Press

There's a duality among Arduino books — they're either appealing to experienced hackers but unfriendly for noobs, or so basic that the reverse is true. John Boxall's 370-page tome attempts to bridge the divide by providing dozens of small and easy projects while serving as a reference on more advanced topics. Each chapter covers a different topic like liquid crystal displays or connecting the Arduino to a GPS receiver. The chapter listings include helpful boldface notations for the actual projects. One downside of the sheer number of projects is that they don't provide a lot of detail. For instance, "Project #40: Building and Controlling a Tank Robot," has readers buy a premade Pololu tank chassis and a motor shield rather than building it themselves, and most of the project is devoted to the sketch.

—*John Baichtal*

Alien Vision: Exploring the EM Spectrum with Imaging Technology

by Austin Richards

$80–$177 SPIE Press

Though Austin Richards is a professional physicist, this book was written for everyone. The introduction covers the basic science clearly, without oversimplfying: light has "colors" we cannot see with our eyes. Chapters focus on different spectral bands: near infrared and ultraviolet, thermal infrared, micro- and millimeter-wave, high energy X- and gamma-rays, plus one on acoustic imaging. The text is well-written and informative, but the highlight is naturally the pictures — especially the comparative pairs, where a normal photograph is presented alongside a hyperspectral image of the same scene. As these imaging tools trickle down into the civilian market, consumer, hobby, and even artistic applications will proliferate. *Alien Vision* gives an inspiring glimpse of just how exciting it's going to be.   —*SMR*

New Maker Tech

ADAFRUIT 16-Channel Servo Shield

$18 adafruit.com

Don't let your Arduino's scarce pulse width modulation outputs crush your hexapod-building dreams! Adafruit's new shield lets you easily control up to 16 servos using just two pins. You can even stack multiple shields to control up to 992 servos through the same pins! Not into bots? It works with LEDs and almost anything else controlled by PWM.

—*Marc de Vinck*

1. Raspberry Pi Camera Board

$25 element14.com

Give your Pi the gift of sight with the official Raspberry Pi Foundation Camera Board. This small, inexpensive accessory connects to the dedicated camera serial interface and delivers 5-megapixel still images or 1080p video directly to the processor. Higher bandwidth than a webcam, it also frees up your USB port for other stuff. The latest version of Raspbian has been updated to take full advantage of this device.

—*Matt Richardson*

2. Arduino Robot

$275 makershed.com

The latest item from Arduino makes a great intro to the world of programmable microcontrollers and robotics. This no-solder kit features two ATmega32u4 microcontrollers, the same chip found on Arduino's Leonardo. Includes color LCD screen, SD card reader, digital compass, speaker, potentiometer, buttons, and more. And there's plenty of space for building your own custom circuits right on the board. A plug-and-play expansion pack is available for makers seeking advanced sensor options.

—*MV*

3. BottleWorks Replicator Refit Kits

$150-$175 bctechno logicalsolutions.com

Known among MakerBot users as "Bottleworks," Bradley Pearce has used his machine-shop chops to turn out some lovely billet-aluminum upgrades for Replicator-series printers. Starting with a set of replacements for the original Replicator's plastic platform arms, Bradley has since introduced upgrade arms for the Replicator 2, as well as a removable glass heated build platform.

All of the upgrades require some disassembly, but initial reports are that they work great.

—*John Abella*

4. Raspberry Pi Wireless Inventors Kit

$78 ciseco.co.uk

Jump fearlessly into the world of wireless Pi! Kit includes both a wireless transceiver daughter board that connects to your Pi and a XinoRF, an Arduino-compatible development board with a built-in transceiver. Use the Raspberry Pi to wirelessly program the XinoRF, then send data back and forth between the two. Create a wireless game controller or a wireless sensor node! Comes with a selection of components to get you started, and a preflashed SD card that boots all the software and libraries ready-to-go.

—*MR*

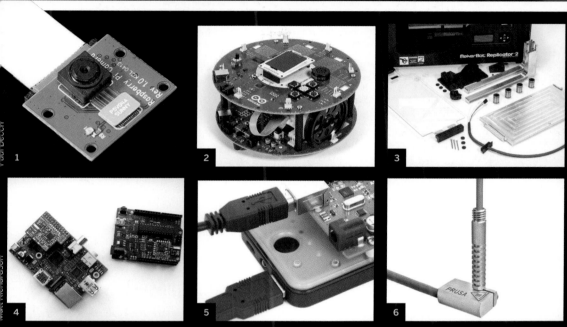

1

2

3

4

5

6

5. MTS Smart Power Base

$70 **smartpowerbase.com**

Looking to run your Arduino, BeagleBone, Netduino, Raspberry Pi, or other compatible-footprint board on battery power? This rechargeable lithium battery pack is a great option. It delivers regulated 5VDC power at up to 1A and can be recharged hundreds of times.

Untether your micro like never before!

—MV

6. Prusa Nozzle

$95 **prusanozzle.org**

Core RepRap developer Josef Průša spent 15 months designing and testing his new printer hot end. Where early nozzles relied on PEEK or PTFE heat barriers, the Prusa Nozzle is a single piece of stainless steel. The interior has been mirror-polished, enabling filament removal at any temperature. It's built in the Czech Republic, hand-finished, certified food-safe, and capable of printing high-melting-point plastics (like polycarbonate) that test the limits of conventional extruder designs.

—JA

7. New Out Of Box Software

Free! **raspberrypi.org**

Typically, the first step for new Pi owners is preparing an SD card with the latest version of the official Linux distro. Until recently, this was a bit of a pain, and a barrier for folks just getting started.

Now, thanks to the Raspberry Pi Foundation's New Out of Box Software (NOOBS), installing Raspbian is much easier. Just extract the .ZIP onto a formatted SD card and it will boot to a menu where you can choose the distribution that suits you best.

—MR

8. Cura Version 13.06

Free! **ultimaker.com**

Anyone who has "sliced" (generated toolpaths for a model) for a hobby 3D printer knows the process can be pretty slow. The last two years have seen typical slicing times fall from hours to minutes, but Ultimaker's new CuraEngine slicing code is a tremendous leap forward. Cura slices models so much faster now that they've actually eliminated the button from the interface and made slicing into a background process. Models that used to take minutes now slice in seconds. Cura can also drive RepRap-style printers.

—JA

New in the Shed

Servos, Sensors, and Sun-Screening Shirts

Finding new products for Maker Shed, our online store of kits, components, and creative inventions (makershed.com), is actually pretty similar to the best part of cracking open the pages of MAKE: It's that moment when the gears begin to turn and we visualize how to apply new tools and techniques to an idea that seemed out of reach just an hour ago. That's how we come across cool new projects to make, and how we look for products to offer in the Shed.

We've just launched more than 50 new items — everything from quadcopters and microcontroller accessories to craft kits and robots. Here are some recent additions we think you'll enjoy.

—*Eric Weinhoffer, Product Dev Engineer*

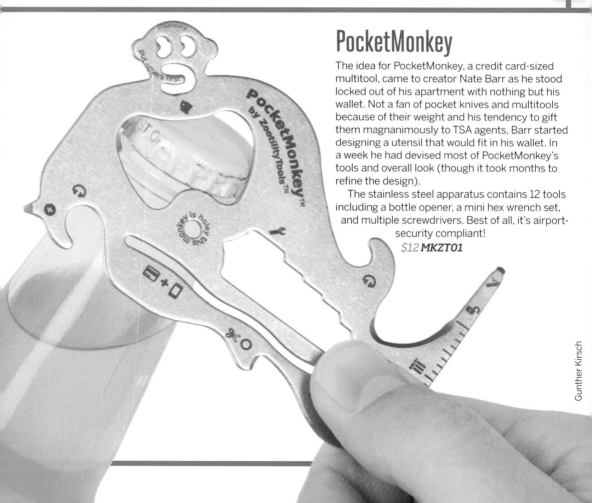

PocketMonkey

The idea for PocketMonkey, a credit card-sized multitool, came to creator Nate Barr as he stood locked out of his apartment with nothing but his wallet. Not a fan of pocket knives and multitools because of their weight and his tendency to gift them magnanimously to TSA agents, Barr started designing a utensil that would fit in his wallet. In a week he had devised most of PocketMonkey's tools and overall look (though it took months to refine the design).

The stainless steel apparatus contains 12 tools including a bottle opener, a mini hex wrench set, and multiple screwdrivers. Best of all, it's airport-security compliant!

$12 **MKZT01**

Gunther Kirsch

Diyode CodeShield

Simon Clark and the good people of the Diyode Community had taught Arduino for a couple of years, but it wasn't until they tried to bring a workshop to Clark's daughter's 4th grade class that they thought about how to streamline their course. The result was creating an Arduino shield packed with LEDs, a switch, servomotor, piezo buzzer, and more.

"We needed to teach code first, and we needed to have successes happening from the moment kids sit down in front of the Arduino," Clark explains. If your LED doesn't blink with this, you can rule out a hardware problem and focus on the programming.
$30 **MKDY1**

Lumi Inkodye

Lumi founder Jesse Genet's inspiration for this specially formulated, light-reactive dye is one that we in the Shed easily relate to: "I was 16 and desperately trying to make cool T-shirts in my parents' basement," Genet says. Her creation, Inkodye, turned out great — just spread it on any natural fiber (like a cotton T-shirt), lay on a negative of your favorite image, put it in the sun, and watch as the image develops before your eyes. The resulting print is permanent and will stand up to washing and even bleach. The photo fidelity is pretty incredible and it's cheaper than a screen-printing kit!
$30 **MKLC01**

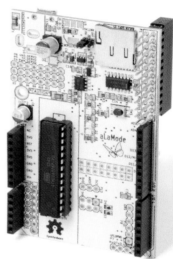

AlaMode

Combining the power of the Raspberry Pi with the versatility of Arduino, this new shield has quickly become one of our favorites. With it, you can run the Arduino software (Arduino IDE) on Linux from the Raspberry Pi. It lets you program the Arduino directly from the Pi and start building complex projects, like a mobile robot that broadcasts to a web page what it sees, hears, or touches.
$35 **MKWY1**

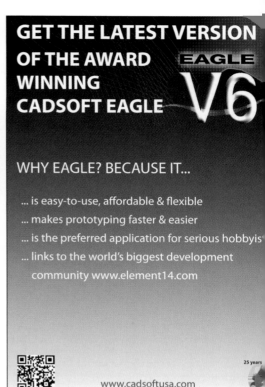

Make: Marketplace

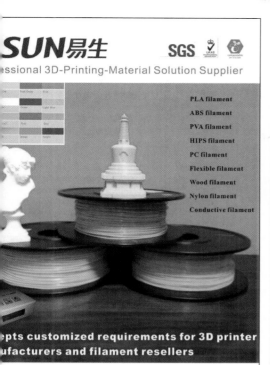

Make: Marketplace

MAKEY

SHANNON WHEELER

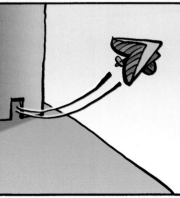